NOURISH AND FLOURISH

This book, along with any associated content or subsequent updates, can be accessed at https://hdl.handle.net/10986/44472.

Scan to see all titles in this series.

GLOBAL WATER MONITORING REPORT

NOURISH AND FLOURISH

Water Solutions to Feed 10 Billion People on a Livable Planet

Amal Talbi, Pieter Waalewijn, Poolad Karimi, IJsbrand De Jong, Francois Onimus, Esha Zaveri, Bogachan Benli, Amadou Ba, Ruyi Li, and Heather Skilling

WORLD BANK GROUP

1 2 3 4   29 28 27 26

ISBN (paper): 978-1-4648-0782-4
ISBN (electronic): 978-1-4648-0783-1
DOI: 10.1596/978-1-4648-0782-4

Cover design: Melina Rose Yingling, Global Corporate Solutions, World Bank Group.

Library of Congress Control Number: 2026905186

Global Water Monitoring Report

The Global Water Monitoring Report series provides regular updates on the state of global water resources and highlights priority areas for investment and policy interventions. Each edition offers an in-depth analysis of a critical water and development issue. This series aims to support the World Bank Water Global Department in its operations, policy dialogue, and engagement with partners.

Titles in This Series

2026

Nourish and Flourish: Water Solutions to Feed 10 Billion People on a Livable Planet, Amal Talbi, Pieter Waalewijn, Poolad Karimi, IJsbrand De Jong, Francois Onimus, Esha Zaveri, Bogachan Benli, Amadou Ba, Ruyi Li, and Heather Skilling

2025

Continental Drying: A Threat to Our Common Future, Fan Zhang, Christian Borja-Vega, Hrishikesh Arvind Chandanpurkar, James Famiglietti, Rick Hogeboom, Regassa Namara, Zarif Rasul, Pavel Luengas-Sierra, and Deyu Rao

All books in the Global Water Monitoring Report series are available for free at https://hdl.handle.net/10986/43724.

Contents

Foreword

There is no food without water. But the defining water challenge is not scarcity—it is imbalance between where water is available and where crops are produced.

In some regions, water is being extracted far beyond what nature can replenish, pushing ecosystems past their limits. In others, abundant water resources remain underutilized, limiting agricultural production and leaving millions without the jobs and food security that better water development could provide.

Nearly half of today's food production already depends on transgressing local natural resources thresholds for water, land, and biodiversity. By 2050, nearly 10 billion people will need to be fed. The challenge we face is not simply to use more or less water but to use water more effectively, at the right time, at the right place, and for the right purpose.

Nourish and Flourish: Water Solutions to Feed 10 Billion People on a Livable Planet makes clear that no single approach provides the optimal solution in all cases. The right choices depend on how much water a country has and the role trade plays in meeting its food needs. This water-food nexus framework is the analytical heart of the report. It offers a practical guide for deciding where rainfed agriculture can be relied on to produce more food, where irrigation should expand to unlock jobs and food production, where water use must be rebalanced to protect ecosystems and future productivity, and where trade offers a more sustainable path for balancing water needs with food security. Water is a core asset not only for food but also for economic and social development. Expanding irrigation alone, within local water thresholds, could create at least 245 million jobs, particularly in Sub-Saharan Africa, where much of this potential remains largely untapped.

Achieving transformation in the way we use our water for food requires fundamental shifts in how agricultural water systems are designed and delivered. Past investments delivered real gains; however, conditions have changed rapidly, and today's realities call for reliable water services that can adapt to change, perform over time, and are supported by capable institutions, clear rules, and better information.

Unlocking the financing needed for this transformation is within reach, not only by mobilizing new resources but also by redirecting existing ones and scaling private investment. The report underscores private sector participation as a critical lever for mobilizing capital, improving efficiency, and accelerating innovation in agricultural water management, from farmer-led irrigation to public irrigation schemes. Enabled by supportive public policies, institutions, and regulations, private capital can help deliver the scale needed—aligned with the World Bank Group's broader efforts to advance agribusiness and job creation through initiatives such as Water Forward and AgriConnect.

Nourish and Flourish also reflects the World Bank Group's renewed approach to how we create, share, and deploy resources through the Knowledge Bank structure, which leverages the International Bank for Reconstruction and Development, International Development Association, International Finance Corporation, and Multilateral Investment Guarantee Agency to deliver a more integrated approach. Transforming agricultural water management depends on data that are accessible, actionable, and fit for decision-making. By leveraging satellite imagery, digital tools, and analytics, this report exemplifies how the World Bank Group can translate global evidence into country-specific choices—embedding knowledge directly into policy design, investment planning, and implementation.

The recommendations of *Nourish and Flourish* are grounded in the World Bank Group Water Strategy. Through its Water for Food and Water for the Planet pillars, the strategy addresses the twin challenge of water and food security by strengthening food production systems and improving farmer livelihoods. It provides the framework for translating knowledge into action, mobilizing the full suite of World Bank Group knowledge, financing, and risk management instruments to support countries in implementing these recommendations.

Feeding the world within environmental limits is not just a question of producing more food but also of making better choices about what we eat, as well as where and how food is produced—and recognizing that water, managed well, is not a constraint but an opportunity in many parts of the world. These choices will determine whether agricultural water supports jobs and growth, strengthens resilience, and protects the natural systems on which food depends. *Nourish and Flourish* supports policy makers in navigating these trade-offs and making those choices more effectively.

Guangzhe Chen
Vice President for Planet
World Bank Group

Acknowledgments

Nourish and Flourish: Water Solutions to Feed 10 Billion People on a Livable Planet was written by Amal Talbi, Pieter Waalewijn, Poolad Karimi, IJsbrand De Jong, Francois Onimus, Esha Dilip Zaveri, Bogachan Benli, Amadou Ba, Ruyi Li, and Heather Skilling. The work was carried out under the leadership and guidance of Guangzhe Chen (Vice President, Planet, World Bank), Juergen Voegele (former Vice President, Planet, World Bank), and Saroj Kumar Jha (Global Director for Water, World Bank). The team also extends its gratitude to Chakib Jenane (Regional Practice Director, Planet, Western and Central Africa Region, World Bank) and the management team of the World Bank Water Global Practice for their valuable advice.

The authors are grateful to a core team of World Bank Group contributors, including Richard Colback, Francois Bertone, Gabriella Izzi, Luis Loyola, Marie-Laure Lajaunie, Xiaokai Li, and Peter Goodman. Core external contributors include Jonathan Denison (consultant), Brian Davidson (formerly University of Melbourne), Aziz Agrebi (Ecole des Mines de Paris), Philippe Renard (University of Neuchatel), Farzad Taheripour (Purdue University), David Blakeslee (New York University Abu Dhabi), Ram Fishman (Tel Aviv University), Karun Verma (Thapar Institute of Engineering and Technology), Dustin Garrick (University of Waterloo), Lucy Annabella Banisch (consultant), Joel Kolker (retired, World Bank), Edoardo Borgomeo (University of Cambridge), Matthew Champness (consultant), Petra Hellegers (Wageningen University), Paul van Hofwegen (retired, World Bank), Tejasvi Hora (University of Waterloo), Ruth Meinzen-Dick (formerly International Food Policy Research Institute), Mark Giordano (Georgetown University), Dan Uriel (Tel Aviv University), and Alisher Mirzabaev (International Rice Research Institute).

An extended team of World Bank and external specialists provided input and expertise. From the World Bank Group, these include Sergiy Zorya, Qiong (Janet) Lu, Regassa Namara, Stuti Sharma, Ranu Sinha, Sean Nelson, Gabriel Seth Sidman, Joop Stoutjesdijk, Zelalem Tesfaye Mekonnen, Greg Browder, David Casanova, Abdulhamid Azad, Wenhe Zhang, Emmanuel Chinedu Umolu, Laura Bonzanigo, Heba Yaken Aref Ahmed,

Alfonso Alvestegui, Sylvestre Bea, Christian Borja-Vega, Abdur Rahim Safi, Willibroad Gabila Buma, Eileen Burke, Anju Gaur, Si Gou, Nagaraja Harshadeep, Aude-Sophie Rodella, Fan Zhang, Jason Lu, Talajeh Livani, Harriet Nattabi, Aissata Delphine Bama Nati, Tarasinta Perwitasari, Satya Priya, David Wanyoike, Ajith Radakarishnan, Imran Sajid, Karamoko Sanogo, Ahmed Shawky, Yitbarek Tessema, Remi Trier, Richard Abdulnour, Dominick Revell de Waal, Marcus J. Wishart, Ousmane Yida Yaya-Bocoum, Persephone Economou, Anders Jagerskog, Sara Datturi, Nathan L. Engle, Chris Fischer, William R. Sutton, Claire Chase, Pierre Jacques Lorillou, and Jakob Andre Eklund. External collaborators include Ezio Todini (Italian Hydrological Society), Paolo Reggiani (consultant), Oleksiy Boyko (consultant), Lorenzo Rosa (consultant), Gustavo Saltiel (retired, World Bank), Wilfried Hundertmark (consultant), David J. Molden (consultant), and Nozilakhon Mukhamedova (consultant).

This report has been enriched by extensive regional consultations involving academia, civil society, government counterparts, practitioners, and development and research partners across the globe. The team sincerely thanks all contributors for their valuable perspectives and engagement. These contributors include Aisha Abdulkadiri (Ahmadu Bello University), K. O. Adekalu (Obafemi Awolowo University), Francesca Battistelli (World Resources Institute), Volodymyr Bilokon (Ministry of Ecology and Natural Resource and State Service of Geology and Subsoil, Ukraine), Helena Cardenas (The Nature Conservancy), Nabil Chemaly (Hilton Foundation), Srinivas Chokkakula (Center of Policy Research, India), Jude Cobbing (Daugherty Water for Food Global Institute, University of Nebraska), Luisa Galvao (WaterAid), Dieter Gerten (Potsdam Institute for Climate Impact Research, Humboldt-Universität zu Berlin), Meike van Ginneken (Netherlands Water Envoy), Maarten Gischler (Netherlands Ministry of Foreign Affairs), Bruno Grawitz (Société du Canal de Provence, France), Xiaogang He (National University of Singapore), Chetan Hebbale (The Nature Conservancy), Nicole Horvath (Xylem), H. E. Igbadun (Ahmadu Bello University), Vincent Kabuti (Kenya Ministry of Water, Sanitation and Irrigation), Vivek P. Kapadia (Water and Climate Lab, IIT Gandhinagar, India), Hans Komakech (Nelson Mandela African Institute of Science and Technology), Nicole Lefore (Daugherty Water for Food Global Institute, University of Nebraska), Wenfeng Liu (China Agricultural University), Florence Deram Malerbe (KER AR ROC'H, France), Emmanuel Manzungu (University of Zimbabwe), Peter McCornick (Daugherty Water for Food Global Institute, University of Nebraska), Rachael McDonnell (International Water Management Institute), Vimal Mishra (Water and Climate Lab, IIT Gandhinagar, India),

Abubakr Muhammad (Supreme Council for Sharia, Nigeria), S. M. Musa (Abubakar Tafawa Balewa University), Kalunga Nkonde (Zambia Ministry of Water Development and Sanitation), Theib Oweis (International Center for Agricultural Research in the Dry Areas), Celine Papin (Eau, Agriculture et Territoires), Lori Pearson (Catholic Relief Services), Enrique Playán (Spanish Research Agency), Tess Russo (Gates Foundation), Lisa Schechtman (Egeria Strategies), Stephane Tromilin (Schneider Electric), Ann Vaughn (CARE), Gert Jan Veldwisch (Wageningen University), Sara Walker (World Resources Institute), William Wallock (Climate Policy Initiative), and Philip Woodhouse (Global Development Institute, University of Manchester). Particular thanks are extended to Bancy Mati for a productive partnership with Jomo Kenyatta University.

The team extends its gratitude to peer reviewers for their invaluable feedback and advice, including Soumya Balasubramanya (Senior Economist, World Bank), Qimiao Fan (Division Director, World Bank), Anup Jagwani (Senior Manager, International Finance Corporation), Hans Komakech (Researcher/Faculty, Nelson Mandela African Institute of Science and Technology, Tanzania), Lifeng Li (Director, Land and Water Division, Food and Agriculture Organization of the United Nations), Wenfeng Liu (Professor, China Agricultural University), Shobha Shetty (Global Director, Agriculture Practice, World Bank), and Winston Yu (Practice Manager, SCAWA, World Bank).

The team thanks the World Bank publishing team, including Cindy Fisher, Jewel McFadden, and Mark McClure. Honora Mara was the copy editor, Catherine Farley was the proofreader, Melina Rose Yingling designed the cover and interior, and Patricio Silva Castillo designed the graphics. We are grateful for the design input of the firm Clarity and the additional editorial support from Bruce Ross-Larson and Elana Bregin. The team appreciates the valuable guidance and support from the World Bank communications team, including Nigina Alieva, Erin Ann Barrett, Ada Calderon, Sarah Farhat, Victoria Ormaechea, Natasha Chantale Skreslet, and Raphaelle Vulliet. The team thanks the World Bank Cartography Unit for their support. Yasmin Angeles supported the team with critical organizational and administrative services.

This publication received the support of the Global Water Security and Sanitation Partnership (GWSP). GWSP is a multidonor trust fund administered by the World Bank Group's Water Department and supported by Australia's Department of Foreign Affairs and Trade; Austria's Federal Ministry of Finance; Denmark's Ministry of Foreign Affairs; the Gates Foundation; the Netherlands' Ministry of Foreign

Affairs; Spain's Ministry of Economy, Trade and Business; the Swedish International Development Cooperation Agency; Switzerland's State Secretariat for Economic Affairs; the Swiss Agency for Development and Cooperation; and the United Kingdom's Foreign, Commonwealth and Development Office.

Main Messages

By 2050, about 10 billion people will need to be fed, yet current water and agrifood systems, including agricultural water management (AWM) practices, can sustainably feed only 3.4 billion people (Gerten et al. 2020). Today's AWM practices lead to imbalanced water resources use—wasting opportunities to enhance water productivity by addressing both overuse and underuse. This report underscores the urgency in addressing AWM shortcomings while helping policy makers navigate the difficult trade-offs related to AWM.

Nourish and Flourish: Water Solutions to Feed 10 Billion People on a Livable Planet introduces a new framing for AWM, categorizing countries on the basis of two key factors: their level of water stress (how abundant or scarce their water resources are) and their food trade position (whether they are net importers or exporters of calories). This water-food nexus framework serves as an entry point for policy makers to navigate the AWM trade-offs based on local realities, helping to tailor strategies that balance food security, water resource sustainability, and economic development. The report provides policy makers with practical examples of successful AWM policies, illustrates implementation pathways, and offers practical strategies for leveraging AWM to deliver economywide benefits. Its assessment of AWM financing needs reveals that the amount, although substantial, falls below current spending by most governments on agriculture support and subsidies. This comparison highlights the opportunity to effectively repurpose existing resources and subsidies.

AWM can enhance productivity, reduce the crop water footprint, and reallocate water to ecosystems or downstream users without compromising food production at the global scale. A 10 percent increase in agricultural productivity can reduce the likelihood of poverty by 2.5 to 3.0 percent (Oseni et al. 2014). The report estimates that sustainable conversion of all estimated rainfed agriculture areas to irrigation, within the limits of local water thresholds, will conservatively create about 245 million jobs—with most of these jobs, about 218 million, in Sub-Saharan Africa.

The forthcoming technical background report will be available at https://hdl.handle.net/10986/44472.

References

Gerten, Dieter, Vera Heck, Jonas Jägermeyr, et al. 2020. "Feeding Ten Billion People Is Possible Within Four Terrestrial Planetary Boundaries." *Nature Sustainability* 3: 200–8.

Oseni, Gbemisola, Kevin McGee, and Andrew Dabalen. 2014. "Can Agricultural Households Farm Their Way out of Poverty?" Policy Research Working Paper 7093, World Bank. https://openknowledge.worldbank.org/server/api/core/bitstreams/fa906994-3691-5df3-95c0-2d7c96b1d2fc/content.

Abbreviations

AI	artificial intelligence
AWM	agricultural water management
COVID-19	coronavirus disease 2019
ET	evapotranspiration
FLID	farmer-led irrigation development
GDP	gross domestic product
ha	hectare(s)
km^3	cubic kilometer(s)
O&M	operation and maintenance
PPP	public-private partnership
SAED	National Company for the Development and Exploitation of the Lands of the Senegal River Delta (Société Nationale d'Aménagement et d'Exploitation des Terres du Delta du Fleuve Sénégal)
SCP	Société du Canal de Provence
SEIASA	State Society for Agricultural Infrastructure
SSA	Sub-Saharan Africa

CHAPTER 1

Introduction

What is new in this report, and why is it important?

Conventional discussion of agricultural water management (AWM) presents a stark trade-off between policies that favor irrigated farming and those that favor the environment; *Nourish and Flourish* demonstrates how AWM can transform both water and food security within local freshwater thresholds.[1]

What is new?

The report is the first global analysis of how irrigation, within the larger context of AWM, can benefit economic, environmental, and human development across different regions, using a data-driven empirical approach with geospatial data and household surveys across 58 countries (Blakeslee et al. 2024). It describes the multiple benefits from a new approach to AWM in terms of impact on three strategic policy objectives: people, prosperity, and planet.[2]

Agricultural water management encompasses the use of two water sources: green water and blue water. *Green water,* soil moisture from rainfall, represents 87 percent of agricultural water use. *Blue water,* which flows in streams, rivers, reservoirs, and aquifers, is abstracted, stored, and distributed for irrigation. Agricultural water management practices range from wholly rainfed systems, to systems that supplement green water with blue, to irrigation systems that mostly or fully depend on blue water. All agricultural water management is to be understood in the wider water cycle and as part of nature-based solutions and circular economy principles.

Nourish and Flourish presents a new framework for policy makers navigating difficult AWM trade-offs and seeking contextually appropriate solutions. Categorizing countries by their level of water stress and whether they are net calorie importers or exporters, the water-food nexus framework provides an analytical tool to help policy makers move from generic one-size-fits-all fixes to country-specific pathways and actions (refer to chapter 3). The report offers readers potential policy actions, selected to maximize benefits or minimize threats depending on where a country is situated within the water-food nexus framework.

By addressing the range of AWM practices—from rainfed agriculture to fully controlled irrigation, from public to private investment, and from traditional knowledge to artificial intelligence (AI)—the framework and report are broadly applicable to all AWM scenarios. The tools and approaches are designed to accommodate the range of AWM conditions, challenges, and objectives, identifying underexploited opportunities for transformation such as the expansion and modernization of irrigation.[3]

The report recognizes that transforming AWM into a scalable, resilient system requires moving decisively toward service-driven, data-enabled irrigation markets. Evidence from India, Morocco, Peru, and Uganda shows that, when risks are shared transparently, revenue streams are predictable and that, when regulation supports cost recovery, private capital rapidly enters—from small and medium enterprises providing installations, operation and maintenance, and digital advisory services to larger firms financing infrastructure through guarantees and blended instruments. Digital innovations such as remote sensing, AI-based groundwater mapping, and hydro-informatics open entirely new investable frontiers that reduce decision-making uncertainty and boost water productivity. With AWM expansion projected to generate more than 200 million jobs, the private sector becomes not only a financier but also the primary engine of employment, service delivery, and technological modernization.

Enabling farmer-led irrigation development (FLID)[4] offers one prominent opportunity to accelerate irrigation expansion (IFC and IFAD 2023; Izzi et al. 2021), including through the power of the private sector. Smallholder farmers represent 84 percent of the world's farmers, use 24 percent of agricultural land, and produce about 30 to 34 percent of the world's food (Ricciardi et al. 2018). Collectively, they have an important impact on food production and the local economy. FLID offers a replicable blueprint for empowering smallholders as agents of food security in the right conditions (Duker 2023), including easily accessible water sources and potential to mobilize farmers' own capital investment.

Farmer-led irrigation development refers to the process whereby farmers, individually or collectively, are the main drivers of improvements and expansion in irrigated agriculture. Farmers—vital private sector participants in the agriculture sector—bring innovation and full or partial investment.

Nourish and Flourish introduces machine learning to identify these high-potential areas (specifically areas with shallow basement aquifers). Such areas offer smallholder farmers access to affordable water sources for irrigation, and the recharge characteristics of this aquifer type (basement crystalline aquifer) make overexploitation unlikely. Box 2.4 in chapter 2 illustrates how machine learning was used to identify nearly 1 million hectares (ha) in the Sahel with sufficient shallow basement groundwater potential to transition from the current rainfed agricultural land to irrigated farming through FLID (World Bank 2025b). The transition creates additional rural economic opportunity because irrigated agriculture typically uses significantly more labor per unit of land than rainfed farming (Wiggins and Lankford 2019).

Despite the transformative potential of tools like remote sensing, AI, and machine learning, these innovative technologies have been underused in advancing AWM. Remote sensing, for instance, can provide near real-time data on crop health, evapotranspiration (ET), crop types and yields, water productivity, and irrigation service delivery performance at a fraction of the time and cost required for conventional census-based data collection. Many agencies, however, still rely exclusively on manual methods. Meanwhile, AI's predictive capabilities offer powerful tools for anticipating climate variability and its impacts, providing policy makers with early warnings and actionable insights, enabling them to better prepare for, and respond to, emerging risks. Modernization programs, such as those in India and Spain, have had transformative impacts through leveraging satellite imagery, AI, and advanced data analytics to deliver near real-time insights into water use, irrigation performance, and overall system efficiency. The use of such technologies has enabled more informed, data-driven decisions for sustainable water management (Ward et al. 2024).

This report establishes that irrigation, within the larger context of AWM, represents a potential powerhouse for job creation. AWM catalyzes a shift from low-value subsistence farming to higher-value market-oriented production and lays the groundwork for labor mobility, agribusiness expansion, and rural industrialization. This transformation generates new employment opportunities in both on-farm and off-farm activities linked to irrigated agriculture and AWM service chains, including a growing role for small and medium enterprises in delivering irrigation equipment, installation, maintenance, and advisory services. Taheripour and Chepeliev (forthcoming) estimate that, in Sub-Saharan Africa (SSA) alone, the scale-up of irrigation would generate at least 218 million jobs—about four jobs for every newly irrigated hectare.

The report further estimates the public investment needs for irrigation, taking into account capital investments needed for conversion from rainfed to sustainable irrigated agriculture, as well as the modernization and regulation of irrigation areas confronted with freshwater resource limitations. Lacking a recent global dataset on unit costs for AWM, the report team conducted a portfolio review of the past 15 years of World Bank–funded AWM projects to extract the unit cost of irrigation and other AWM interventions. The review developed two investment scenarios: (1) a low-investment scenario with 40 percent of currently unsustainable irrigation area modernized and 50 percent of current rainfed area converted to irrigation, and (2) a high-investment scenario with 80 percent of currently unsustainable irrigation area modernized and 100 percent of current rainfed area converted to irrigation. Estimated investment needs were calculated for each scenario by multiplying potential areas for conversion or modernization by unit costs. By estimating investment needs, the report makes it possible to compare needs against current budget support for agriculture. This comparison reinforces conclusions from earlier studies that current producer and consumer subsidies could be partially redirected to finance AWM (OECD 2023).

Why is this report important?

To feed 10 billion people on a livable planet by 2050, the world needs to transform how it manages water in agriculture. Current food production is based on unsustainable imbalances, including the overuse and underuse of water. In an increasingly volatile world, AWM offers an essential risk management strategy that buffers against supply disruptions, such as those experienced during the COVID-19 pandemic and the Russian Federation's invasion of Ukraine, as well as progressive land degradation and climate shocks. Countries that transform AWM will improve livelihoods, boost crop yields, reduce pressure on natural resources, and build resilience to food crises, helping promote long-term economic stability.

Nourish and Flourish underscores the critical link between data and transforming AWM, emphasizing that the challenge is not primarily a lack of data but rather a lack of data that are fit-for-purpose, accessible, and contestable and that integrate the many disciplines involved in impact-accountable AWM. The lack of accurate and accessible data undermines AWM analysis and policy efforts, because critical hydrological, agronomic, socioeconomic, and financial data remain incomplete, fragmented, or outdated. Institutions often lack accountability and quality control mechanisms and have no incentives for data-sharing across agencies. Inaccurate, unreliable, and incomplete data erode stakeholder trust in available information. The result is a cycle in which poor data quality discourages use and the low level of use discourages the collection of data and quality checking. Without reliable information to guide AWM, countries face greater risks to livelihoods, ecosystems, and public budgets—and communities have fewer opportunities to shape and challenge the choices that affect them.

AWM's success requires a country-level reassessment and rebalancing of water security, food security, and food self-sufficiency goals within the global context. Although all countries aspire to ensure that everyone has sufficient and nutritious food, each country must assess its own circumstances and work with citizens and stakeholders to decide how to produce and regulate food and how to prioritize ecologically sound and sustainable practices. Balancing policy priorities requires a clear understanding of the contextual realities—within a country, within a region, and globally.

The challenge: Feeding 10 billion people sustainably

Current water and agrifood systems including present AWM practices can support sustainable food production for only about 3.4 billion people—less than half the world's population of 8 billion people (Gerten et al. 2020). By 2050, an estimated 10 billion people will need to be fed. Without a transformative shift in AWM, deficiencies and inequities in distribution, nutrition, and sustainability will worsen, making it highly unlikely that the additional 2 billion people can be sustainably fed. Even if 3.4 billion people remain sustainably fed under current AWM practices, that number will represent only one-third of the world's population by 2050.

Sustainably feeding a population entails accessing or producing food in a way that does not deplete natural resources and that can be supported by adequate and reliable financial flows. Global agriculture puts heavy pressure on four interlinked planetary boundaries (biosphere integrity, land-system change, freshwater use, and nitrogen flow); almost half of current global food production depends on planetary boundary transgressions.

Although global calorie production is currently adequate to feed the entire world's population (Krishna Bahadur et al. 2018), present AWM assessments reveal dangerous water imbalances. Some regions, such as parts of SSA, underuse their water resources; others, such as parts of South Asia, overexploit theirs. These extremes threaten food security and environmental stability, especially as water shocks intensify. Droughts and floods already devastate agricultural production, destroy livelihoods, and drive up emergency spending.

Current agricultural subsidies often favor foods with lower nutritional value, such as sugar, and subsidy allocations do not prioritize water-smart foods. What we eat matters as much as where and how we grow our food; food systems and AWM are intrinsically connected as farmers change their crop choices and practices in response to the demands of consumers and the food system. For instance, farming and meat production often depend on production systems with low water efficiency: raising livestock with irrigated forage crops using blue water is very water inefficient (20 times less efficient in calories and 6 times less efficient in proteins than plant-based alternatives); by contrast, livestock that extensively graze pastures grown with green water (rainfall) make otherwise unproductive pastureland productive. Meaningful blue and green water impacts as well as potential water savings derive from changes in global diets.

Inclusive economic development requires urgent action in AWM, driving job creation. AWM investments can generate millions of jobs. For instance, converting all rainfed agricultural land with sustainable blue water access to irrigation could create at least 245 million jobs, with nearly 90 percent of these jobs—about 218 million—created in SSA (Taheripour and Chepeliev, forthcoming). This significant opportunity makes development of AWM in SSA critical for food production, jobs, economic development, and environmental sustainability. Conversely, green water shocks trigger massive job losses in agriculturally dependent areas (Khan et al. 2024). AWM also serves as a powerful tool to reduce inequality, especially for vulnerable smallholders, women, and youth who bear the brunt of climate shocks while lacking access to resources.

Feeding the world's growing population will require US$600 billion to US$1.8 trillion between 2025 and 2050, or US$24 billion to US$70 billion per year globally (Li et al., forthcoming). Data from 90 countries show that total spending on agricultural support reached US$663 billion in 2023.[5] Approximately US$490 billion represented public expenditure, and only about US$27 billion of this amount was spent on irrigation. In 2023, about three-quarters of the US$490 billion in public expenditure was used for agriculture subsidies or payments for production of specific commodities—highly distortive to both production and consumption decisions.

Repurposing existing agricultural financing and subsidies could reduce the need for additional AWM financing (Damania et al. 2023). Significant opportunities exist to repurpose the current spending on harmful subsidies and agricultural support to more effective irrigation interventions, thus engendering greater value for money (World Bank 2025a).

AWM benefits extend far beyond farm-level interventions

As noted earlier, AWM encompasses a spectrum of practices that harness two fundamental water sources: green water and blue water. *Green water* is soil moisture from rainfall that plants use, and it represents 87 percent of agricultural water use. *Blue water* flows in streams, rivers, reservoirs, and aquifers; this water is mobilized (abstracted, stored, and distributed) for irrigation. AWM practices range from wholly rainfed systems to irrigated systems that supplement green water with blue water, to advanced irrigation systems that provide nearly complete water control (figure 1.1). This spectrum reflects flexible interventions tailored to local conditions and climate change impacts.

FIGURE 1.1 The continuum of agricultural water management

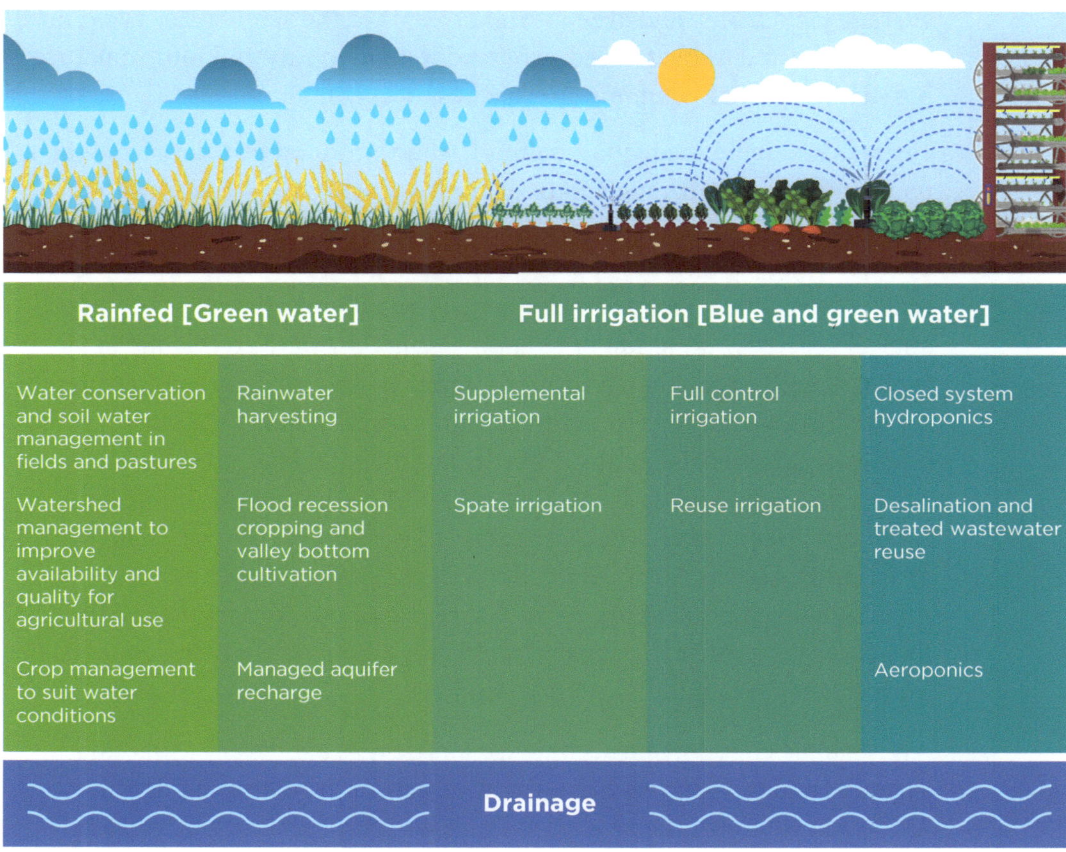

Source: Adapted from Molden 2007.

Each year, about 119,000 cubic kilometers (km^3) of precipitation fall on land; over 60 percent (approximately 74,000 km^3) of this precipitation becomes green water and almost 40 percent (approximately 45,000 km^3) becomes blue water. Humans withdraw about 3,800 to 4,000 km^3 of water per year, or about 9 percent of available blue water, and use 5 to 6 percent for irrigation. More than half of this amount returns to blue water as return flows, with the bulk of actual consumption occurring through ET (global moisture recycling). Blue water consumption is highly visible (and contested) but is relatively small in global terms; green water is the dominant, but hard to manage, driver of crop growth and agricultural water consumption (Global Commission on the Economics of Water 2024).

Applying blue water reduces the timing risks from unpredictable rainfall and increases water productivity in areas with limited or erratic rainfall and green water availability. Blue water is the resource that can be controlled, creating both trade-offs and opportunities for greater impact. Crops need water at specific growth stages. Rainfall cannot be controlled, but blue water irrigation ensures that crops receive water when needed. Using blue water can prevent crop failure during critical periods and boost average yields. Blue and green water are part of the same water cycle, and their fluxes and impacts need to be understood jointly and inform nature-based and circular economy approaches to AWM.

In areas receiving inadequate rainfall, water productivity increases when using blue water. Although counterintuitive, using more (blue) water through irrigation dramatically increases the efficacy of green water uses by increasing productivity per unit of water used, resulting in less water (and land) needed overall for the same level of food production. A meta-analysis of rice production in 12 countries in SSA showed that mean yields varied from 0.6 ton/ha to 2.3 ton/ha under rainfed agriculture and from 2.5 ton/ha to 5.6 ton/ha under irrigation. Irrigated land (AWM that uses blue water) represents only 6.5 percent of the total land used for agriculture (all crops and livestock) and 22.7 percent of the arable land (nonpermanent crops), yet it produces 40 percent of the food (Lankford et al. 2016), demonstrating irrigation's considerable impact on food security. Only 3 percent of the global area equipped for irrigation is in low-income countries, compared to 18 percent in high-income countries; most is located in middle-income countries.[6]

AWM shapes not only farms but also landscapes and river basins, and agricultural land can serve as a water buffer for droughts and floods. AWM can balance intensive agriculture with ecosystem conservation.

Beyond the farm boundary, AWM practices that increase infiltration, reduce surface runoff, and promote biomass growth can enable the reversal of land and rangeland degradation and support biomass accumulation. In highly stressed basins, real water conservation through appropriate policy, regulation, and technology can improve water quality and downstream water availability. Less sedimentation, more stable base flows, and better (local and national) management would then aid wetland ecosystem restoration. Agricultural water needs and allocations can also serve as a flexible water buffer in strategic planning; adjustments in water allocations during drought can reallocate water from agricultural land to critical habitats, cities, industry, or other uses. During floods, agricultural lands can absorb excess water, thus protecting more valuable built assets and lives. In both cases, it is economically prudent for governments or other users to remunerate farmers for providing these environmental services. Such payments can happen at the local, basin, or transboundary level.

AWM offers significant potential for reducing energy-related emissions, reducing methane emissions, and increasing carbon sequestration in support of climate adaptation and mitigation. Irrigation, in particular, is responsible for over 15 percent of agriculture-related greenhouse gas emissions, and groundwater pumping accounts for nearly 90 percent of irrigation's energy use (Qin et al. 2024). Energy audits, electrification, and distributed renewable energy–powered irrigation systems can all reduce emissions and improve revenue, and therefore represent a win-win if awareness is raised and access to credit is addressed. Rice, the most widely grown irrigated crop in the world, is a major contributor to methane emissions. Practices such as alternate wetting and drying can reduce methane emissions by up to 53 percent (Jiang et al. 2019).[7] They can also provide water savings, but realizing these savings requires major reengineering, financial incentives, and governance changes for large-scale rice irrigation areas. Irrigation, like forests, globally increases terrestrial ET, the combined process through which water moves from the land surface back into the atmosphere, encompassing both evaporation and transpiration. Through ET, irrigation contributes to atmospheric moisture recycling and reduces local heat stress. This effect benefits between 0.79 billion and 1.29 billion people, mostly in South Asia (Thiery et al. 2020). Irrigation is a major, underrecognized moderator of extreme heat, but the future of these benefits depends on sustainable future AWM models.

Notes

1. This report uses "freshwater" and "blue water" interchangeably.
2. The report uses "people, prosperity, and planet" to categorize the benefits delivered by AWM—such as improved livelihoods and off-farm employment; increased agricultural GDP, employment, and productivity; reduced land and water pressures; and climate adaptation and mitigation (refer to figure 2.1 in chapter 2).
3. "Irrigation expansion" refers to the conversion of rainfed agricultural land to irrigated agriculture within the local water threshold. It does not imply an expansion of the land area used for agriculture.
4. World Bank Group, "Farmer-led Irrigation Development (FLID)," https://www.worldbank.org/en/topic/water/brief/farmer-led-irrigation-development-flid.
5. Data from Global Alliance for Food Security's Global Food and Nutrition Security Dashboard, https://www.gafs.info/home/. The estimate of US$663 billion includes US$175 billion of market price support, a transfer from consumers to farmers as a result of import protection. This part of the support is not readily available for governments to repurpose.
6. Data from FAOSTAT, "Land Use," Food and Agriculture Organization of the United Nations (accessed August 19, 2025), https://www.fao.org/faostat/en/#data/RL.
7. Refer also to Intergovernmental Panel on Climate Change, "Sixth Assessment Report," https://www.ipcc.ch/assessment-report/ar6/.

References

Blakeslee, David, Rema Fishman, Tanuj Hora, and Esha Zaveri. 2024. "The Role of Irrigation in Local Development: Evidence from Around the World." Background paper for *Nourish and Flourish: Water Solutions to Feed 10 Billion People on a Livable Planet*. World Bank.

Damania, Richard, Esteban Balseca, Charlotte de Fontaubert, et al. 2023. *Detox Development: Repurposing Environmentally Harmful Subsidies*. World Bank. http://hdl.handle.net/10986/39423.

Duker, Annelieke. 2023. "Viewpoint—Seeing Like a Farmer: How Irrigation Policies May Undermine Farmer-Led Irrigation in Sub-Saharan Africa." *Water Alternatives* 16: 892–99.

Gerten, Dieter, Vera Heck, Jonas Jägermeyr, et al. 2020. "Feeding Ten Billion People Is Possible Within Four Terrestrial Planetary Boundaries." *Nature Sustainability* 3: 200–8.

Global Commission on the Economics of Water. 2024. *The Economics of Water: Valuing the Hydrological Cycle as a Global Common Good*. Global Commission on the Economics of Water. https://economicsofwater.watercommission.org/report/economics-of-water.pdf.

IFC (International Finance Corporation) and IFAD (The International Fund for Agricultural Development). 2023 *Handbook for Scaling Irrigation Systems*. IFC and IFAD. https://www.ifc.org/content/dam/ifc/doc/mgrt/2023000867engeng001-web-pdf-optimized.pdf.

Izzi, Gabrielle, Jonathan Denison, and Geert Jan Veldwisch, eds. 2021. *The Farmer-Led Irrigation Development Guide: A What, Why and How-To for Intervention Design*. World Bank.

Jiang, Yu, Daniela Carrijo, Shan Huang, et al. 2019. "Water Management to Mitigate the Global Warming Potential of Rice Systems: A Global Meta-Analysis." *Field Crops Research* 234: 47–54. https://doi.org/10.1016/j.fcr.2019.02.010.

Khan, Amjad M., L. Kuate, R. Pongou, and F. Zhang. 2024. "Weather, Water, and Work: Climatic Water Variability and Labor Market Outcomes in Sub-Saharan Africa." Policy Research Working Paper 10823, World Bank.

Krishna Bahadur, K. C., Goretty M. Dias, Anastasia Veeramani, et al. 2018. "When Too Much Isn't Enough: Does Current Food Production Meet Global Nutritional Needs?" *PLoS ONE* 132018: e0205683.

Lankford, Bruce, Ian Makin, Nathanial Matthews, Peter G. McCornick, Andrew Noble, and Tushaar Shah. 2016. "A Compact to Revitalise Large-Scale Irrigation Systems Using a Leadership-Partnership-Ownership 'Theory of Change.'" *Water Alternatives* 9 (1): 1–32.

Li, Ruyi, Francois Onimus, Poolad Karimi, et al. Forthcoming. "Portfolio Review of World Bank Irrigation Project, Unit Costs and Total Financing Need." Background note for *Nourish and Flourish: Water Solutions to Feed 10 Billion People on a Livable Planet*. World Bank.

Molden, David, ed. 2007. *Water for Food, Water for Life: A Comprehensive Assessment of Water Management in Agriculture*. Earthscan; International Water Management Institute.

OECD (Organisation for Economic Co-operation and Development). 2023. *Agricultural Policy Monitoring and Evaluation 2023: Adapting Agriculture to Climate Change*. OECD Publishing. https://doi.org/10.1787/b14de474-en.

Qin, Jingxiu, Weili Duan, Stefan Zou, et al. 2024. "Global Energy Use and Carbon Emissions from Irrigated Agriculture." *Nature Communications* 15: 3084. https://doi.org/10.1038/s41467-024-47383-5.

Ricciardi, Vincent, Navin Ramankutty, Zia Mehrabi, Larissa Jarvis, and Brenton Chookolingo. 2018. "How Much of the World's Food Do Smallholders Produce?" *Global Food Security* 17: 64–72.

Taheripour, Farzad, and Maksym Chepeliev. Forthcoming. "Assessing Economic Impacts of Changes in Irrigated Area: A Global Analysis Using a Multi-Region Input-Output (MRIO) Framework Based on GTAP 2017 Data Base." Background paper for *Nourish and Flourish: Water Solutions to Feed 10 Billion People on a Livable Planet*. World Bank.

Thiery, Wim, Auke J. Visser, Erich M. Fischer, et al. 2020. "Warming of Hot Extremes Alleviated by Expanding Irrigation." *Nature Communications* 11 (1): 290.

Ward, Christopher, Charles Burt, Svetlana Valieva, Ahmed Shawky, David Casanova, and David Meerbach. 2024. "Innovation and Modernization in Irrigation and Drainage: A Guide to Why, What, and How." World Bank.

Wiggins, Steve, and Bruce Lankford. 2019. "Farmer-Led Irrigation in Sub-Saharan Africa: Building on Farmer Initiatives." Briefing/Policy Paper, ODI Global. https://odi.org/en/publications/farmer-led-irrigation-in-sub-saharan-africa-building-on-farmer-initiatives/.

World Bank. 2025a. *Repurposing Agricultural Support: 2025 Policy Compendium*. World Bank.

World Bank. 2025b. "Sahel Irrigation Strategy." World Bank. http://hdl.handle.net/10986/43750.

CHAPTER 2

Rethinking Agricultural Water Management

Investments in agricultural water management deliver against three strategic policy objectives: People, prosperity, and planet

Investing in agricultural water management (AWM) unlocks multiple development dividends (figure 2.1) against three high strategic development goals: people, prosperity, and planet. AWM aids *people* by supporting better and more stable livelihoods for farmers, raising farmer and farm labor income, and boosting off-farm employment through inclusion of more diverse and equitable layers of society, including genders. AWM contributes to *prosperity* by boosting agricultural GDP, stabilizing food prices, and increasing productivity across food systems. AWM supports the *planet* by reducing land and water pressures and supporting climate adaptation and mitigation.

Three shifts needed to transform AWM

Feeding 10 billion people on a livable planet will require rethinking AWM. The following three foundational paradigm shifts are needed to maximize AWM benefits.

Shift 1: Embrace AWM complexity to balance people, prosperity, and planet

The challenge: Current AWM policy often prioritizes quick wins and one-size-fits-all approaches. Policy interventions must envision AWM as a platform for sustainable development, not just a means to boost production. This change requires connecting water, agriculture, environment, and economic growth. Successful AWM interventions depend on navigating local water availability, agricultural systems, institutions, and socioeconomic conditions. Planning frameworks should be capable of managing uncertainty, mediating competing interests, and adapting through iterative learning over time (Waalewijn et al. 2020).

FIGURE 2.1 AWM delivers against three strategic policy objectives: People, prosperity, and planet

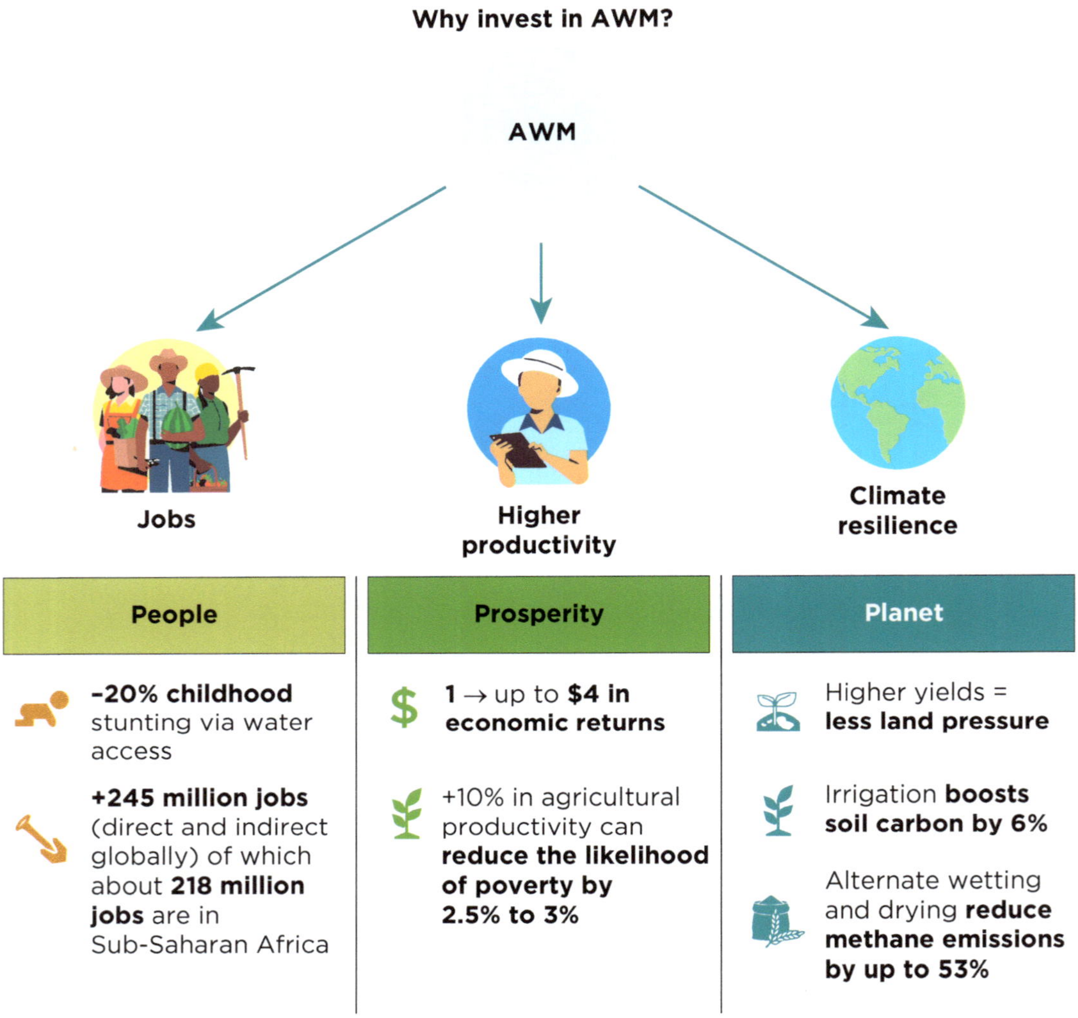

Sources: Rodella et al. 2023 (childhood stunting); Giordano et al. 2023 (economic returns); Qin et al. 2024 and Intergovernmental Panel on Climate Change, "Sixth Assessment Report," https://www.ipcc.ch/assessment-report/ar6/ (methane emissions); Jiang et al. 2019 (methane emissions); Emde et al. 2021 (soil carbon); Taheripour and Chepeliev, forthcoming (job creation); Oseni et al. 2014 (poverty reduction); Damania et al. 2023 (land pressure).
Note: AWM = agricultural water management.

How to make the shift happen: Establish or strengthen multistakeholder platforms, at basin or regional levels, that facilitate structured cross-sectoral dialogue by bringing together farmers, industry, environmental groups, and government agencies to understand local contexts, weigh trade-offs, and codevelop locally accepted solutions (Waalewijn et al. 2020). This process builds shared ownership of decisions, leading to more equitable, sustainable, and politically durable water and land use arrangements. Kenya and the US state of Nebraska provide good examples of this shift (Ministry of Water, Sanitation and Irrigation, Kenya 2025; Sixt et al. 2019; refer to box 2.1).

BOX 2.1
Achieving transformational shift through stakeholder collaboration

Kenya's National Irrigation Sector Investment Plan represents a fundamental shift from fragmented infrastructure efforts to a coordinated, farmer-centered approach. The plan, cocreated with broad participatory citizen engagement, organizes all government investment, policy, and partner funding along five pathways for action. By targeting high-return agricultural products such as horticulture, and combining investments with technical support, financial services, and market access, the plan prioritizes income gains and resilience for all farmers—especially smallholders. Doing so ensures that irrigation investments translate into inclusive, long-term development.

Nebraska's Natural Resources Districts system includes 23 locally governed bodies bringing together farmers, technical experts, and community representatives to manage groundwater across watershed boundaries. Supporting nearly 100,000 irrigation wells with advanced center-pivot technology, Nebraska leads the United States in irrigated area while maintaining stable groundwater levels. Each district develops tailored solutions based on local conditions: For instance, some focus on stream depletion and others on water quality or localized depletion. The system balances agricultural productivity, long-term sustainability, and interstate water obligations through adaptive management. Nebraska's Natural Resources Districts have evolved from service districts to sophisticated regulatory bodies that impose allocations, fund infrastructure, and coordinate across jurisdictions.

Shift 2: From hardware to services—start small, learn fast, and transform at scale quickly

The challenge: Disconnected projects and infrastructure-centric approaches dominate AWM investment. Many initiatives are designed as isolated projects that do not address underlying system failures such as limited line of sight to higher-level objectives, poor accountability, weak financial sustainability, and fragmented mandates (Waalewijn et al. 2020). Without addressing these foundational weaknesses, investments fail to achieve their desired impact.

How to make the shift happen: Government plays a critical role in creating clear, transparent, and accountable operating rules for irrigation providers and standards for irrigation service. To date, limited regulation of irrigation, whether through a water services regulator or by contract, has led to a lack of standardization of irrigation user charges and fees and limited focus on cost recovery and financial sustainability of irrigation service. Regulatory strengthening builds performance-driven and customer-centric AWM systems that deliver reliable services to farmers with a focus on operation and maintenance (O&M), not just construction. Several encouraging examples exist of greater accountability and regulation in irrigation (Waalewijn et al. 2020).

For instance, the Maharashtra Water Resources Regulatory Authority, as the first independent statutory water regulatory authority in India, adopted a transparent approach to setting the bulk water tariffs for irrigation, domestic, and industrial uses based on cost recovery of operation and maintenance charges.[1] Similarly, in the country of Georgia, the National Energy and Water Supply Regulatory Commission was assigned the added mandate of regulating Georgian Amelioration, the agency responsible for irrigation systems and for water delivery to aquaculture and hydro firms (Asian Development Bank 2016). The commission issued Irrigation Water Supply Rules in 2023, for a three-year period, followed by a new tariff structure in 2025 that calls for 100 percent O&M cost recovery. The commission differentiated the tariff for members of water users organizations that are responsible for secondary and tertiary O&M. These customers pay a reduced tariff reflecting only the costs of the main canal O&M. The price reduction created an incentive for Georgian Amelioration customers to create water users organizations. These regulatory interventions, integrated with institutional reorganization and modernization of volumetric water delivery measurement by installing water meters at the intake and discharge points from a dam or river, support 100 percent recovery of O&M costs for irrigation service delivery.

The inclusion of results-based financing in investment programs, aiming for regulatory strengthening and performance-based contracts, offers tactically important instruments to align incentives across institutions and funding streams, prompting improved efficiency and sustainability of water service delivery. Regulatory reforms both initiate and enable the transition to professionalized service providers and, when combined with the operators' organizational realignment and with modernization investments, create more predictable revenue flows and clear accountability mechanisms. Mali's Office du Niger, for example, introduced a service contract, the Contrat Plan, making water delivery obligations and fee structures explicit, thus adopting a service and performance approach, with formalized accountability metrics and processes (World Bank 2021; box 2.2).

BOX 2.2

Achieving transformational shift through institutional and regulatory change

India's Maharashtra Water Resources Regulatory Authority is one of the few regulators that sets bulk water tariffs for irrigation in addition to domestic and industrial water tariffs; more recently, in 2023, the country of Georgia assigned the National Energy and Water Supply Regulatory Commission responsibility for regulating Georgian Amelioration, the agency responsible for irrigation systems in addition to its regulatory oversight of energy, water supply and sanitation, irrigation, aquaculture, and some hydropower.

In Mali's Office du Niger, the Contrat Plan established a joint planning and accountability mechanism among the government, the irrigation agency, and farmers. This system-level agreement links investment, maintenance, and performance monitoring—shifting the focus from infrastructure delivery to sustained service quality through shared responsibility and institutional coordination.

Collaborative efforts of government, farmers, and suppliers enabled a one-year farmer-led irrigation development pilot program in Uganda to quickly expand to the national level using established technologies and an improved rollout process.

A new generation of AWM policies and interventions may, in fact, not invest in infrastructure at all. The rise of distributed renewable energy solutions and the push for inclusive, fiscally responsible, and fast results are driving a complementary portfolio of investments that shift from building projects to enabling the market ecosystem needed for the widespread adoption of products (for example, solar pumps, microirrigation technologies, and weather and irrigation advisories). Consequently, questions of planning, targeting, financial incentives, private sector–led growth, financial access, de-risking, and guarantees now enter a space traditionally dominated by engineers. Policy and delivery approaches increasingly follow the adage "start small, learn fast (across boundaries), and transform at scale" through coordinated regional, intersectoral programs.

Finally, rapid scaling requires feedback loops from all partners and reflexive management to accelerate improved implementation, supported by enabling policies and public, private, and community partnerships (for more on the role of the private sector, refer to box 2.3) (Davidson et al., forthcoming). The Uganda example of farmer-led irrigation development provides lessons learned for further scaling up in the country and beyond (Denison and Christen 2025).

BOX 2.3
The private sector's role in scaling AWM

The vital role of the private sector in scaling agricultural water management (AWM)—through investment, expertise, and innovation—is increasingly evident. Private sector innovation is critical to filling data gaps, enabling targeted and productive investment, and private expertise drives performance improvements in AWM. To advance private sector engagement, governments and their partners are developing competitive markets for services, including operation and maintenance, delivered by private irrigation operators. Although large-scale private sector players have been successfully engaged in countries including Jordan, Morocco, and Peru, smallholder farmers represent a growing private finance force, particularly through farmer-led irrigation development. Experience with farmer-led irrigation development demonstrates that, when investments are de-risked, access to appropriate finance increase, viable private offtake markets deepen, and small farmers are keen to invest in irrigation. AWM scales and succeeds when it operates as a service, grounded in cost-recovery principles and accountable operators.

Shift 3: End the guesswork—collect the data and use them!

The challenge: AWM decisions often rely on fragmented, outdated, or siloed information. Critical data on water availability, use, infrastructure performance, and environmental impacts remain incomplete, unreliable, or inaccessible across agencies. This lack of data makes it harder for governments to justify priorities, demonstrate value for money, or ensure that investments reflect community needs—ultimately weakening both accountability and outcomes.

How to make the shift happen: Agricultural actors need to institutionalize the development and use of water informatics, focusing on key indicators like evapotranspiration (ET) to gain an accurate understanding of water use and productivity. Leveraging low-cost digital solutions and existing data sources, and making them publicly accessible, can unleash creativity and accountability. Piloting concrete decision-making based on transparent data will provide near real-time insights into water use and productivity.

China's multidecade journey toward ET-based water management shows the power of this approach. Using remote sensing to analyze basin-scale water balance, China has established sustainable water use caps in specific areas and implemented crop pattern adjustments and ET-based water rights and pricing to meet reduction targets. This journey achieved measurable water savings within ecological limits (Wang et al. 2020).

In Uganda, the Micro-Scale Irrigation Program modeled digital integration, using the IrriTrack app to register farmers, give tailored technical and financial advisories, record farm visits, track applications, and record the installation of equipment. Linked to a national management information system, IrriTrack helps monitor and evaluate progress and ensures a consistent approach. More than 1,975 local staff such as engineers and extension practitioners were trained through online modules to support farmers in accessing and maximizing the program and to improve accountability of service delivery (Denison and Christen 2025).

Box 2.4 illustrates the potential for artificial intelligence and machine learning to support the conversion from rainfed agriculture to irrigation in the Sahel, which could lower costs and unlock opportunities for the private sector (Agrebi et al. 2025; World Bank 2025).

BOX 2.4

Machine learning to identify irrigation opportunities in the Sahel

The Sahel region faces unique challenges in securing food and achieving sustainable growth despite freshwater availability, albeit with significant fluctuations across time and space. The region's ecological and socioeconomic fragility, combined with the fact that the agriculture sector accounts for 40 percent of the region's gross domestic product and serves as its largest employer, means that the Sahel is among the world regions most at risk from water shocks.

In five Sahelian countries, a machine learning model identified and mapped shallow groundwater in basement crystalline aquifers. This data-driven approach identified nearly 1 million hectares of land, as illustrated in map B2.4.1, suitable for low-cost, farmer-led irrigation—ending the guesswork and significantly reducing the cost of converting these areas to rainfed agriculture, until now a major constraint in irrigation development in Sub-Saharan Africa. Highlights include the following:

- The activity supports the Sahel Irrigation Strategy in a context where 97 percent of Sahelian agriculture depends on rainfall.
- With support from the University of Neuchâtel, Switzerland, the artificial intelligence tool was trained with thousands of data points on the depth of groundwater—pulled from historical records.
- Mapping confirms significant shallow basement groundwater, well suited for small-scale private irrigation and community scheme development. Sixty-two percent of Sub-Saharan Africa is underlain by relatively shallow water tables within crystalline basement aquifers.
- New World Bank–funded programs are piloting farmer-led irrigation development making use of the mapped data, and optimization of the artificial intelligence model is under way through Geostack Inc. and the Institute National de la Recherche Scientifique of Canada.
- A planned smartphone app will increase model accuracy through user-collected data and verification, while making data more readily available to farmers for decision-making.

(continued)

BOX 2.4
Machine learning to identify irrigation opportunities in the Sahel *(continued)*

MAP B2.4.1 Shallow basement groundwater in the Sahel identified through a machine learning model

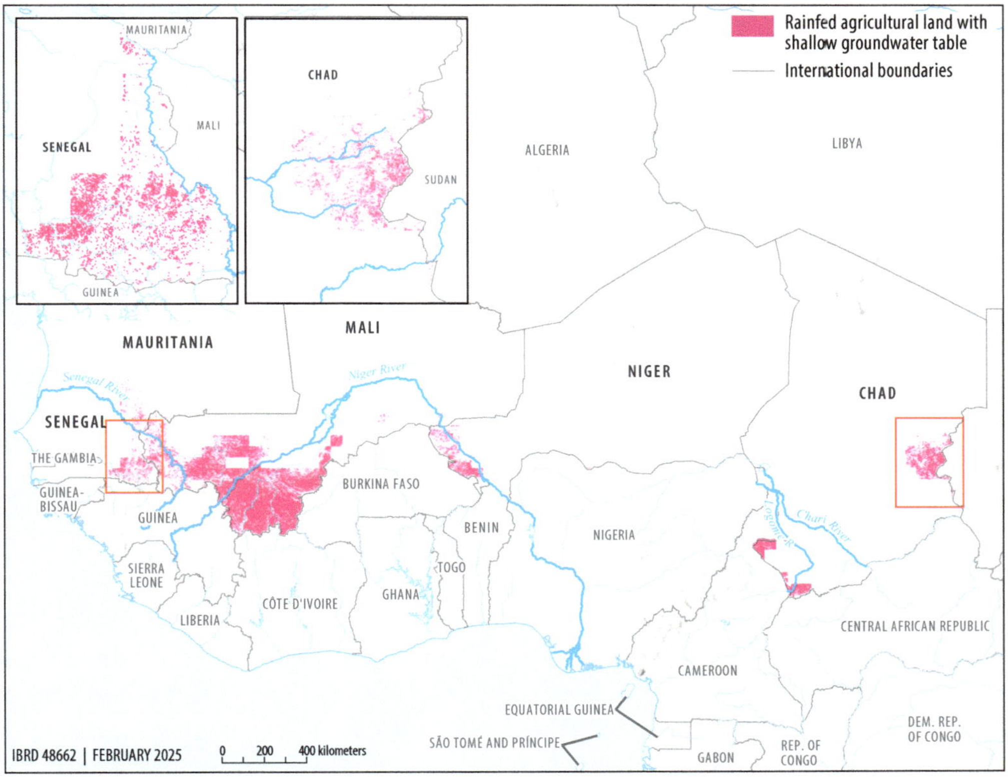

Source: World Bank 2025.

Note

1. Centre for Policy Research, "Know Your Regulator: Special Edition on Water Authorities in Maharashtra and Punjab," https://cprindia.org/know-your-regulator-special-edition-on-water-authorities-in-maharashtra-and-punjab/.

References

Agrebi, Aziz, Francois Bertone, and Philippe Renard. 2025. "Assessment of Static Water Levels in Basement Aquifers Using Machine Learning and Quantification of the Corresponding Irrigation Expansion Potential: A Proof of Concept." Background note for *Nourish and Flourish: Water Solutions to Feed 10 Billion People on a Livable Planet*. World Bank.

Asian Development Bank. 2016. *Second Asian Irrigation Forum: Key Messages and Outcomes*. Asian Development Bank.

Damania, Richard, Esteban Balseca, Charlotte de Fontaubert, et al. 2023. *Detox Development: Repurposing Environmentally Harmful Subsidies*. World Bank. http://hdl.handle.net/10986/39423.

Davidson, Brian, Petra Hellegers, Amal Talbi, Pieter Waalewijn, Poolad Karimi, and Ruyi Li. Forthcoming. "Making Decisions on Agricultural Water Management: A Pathway Approach." Background Paper for *Nourish and Flourish: Water Solutions to Feed 10 Billion People on a Livable Planet*. World Bank.

Denison, Jonathan, and Evan Christen. 2025. "Learning Note on the Design and Rollout of the Uganda Micro-Scale Irrigation Program: Enabling Farmer-Led Irrigation Development by Closing the Affordability and Knowledge Gap." World Bank.

Emde, David, Kirsten D. Hannam, Ilka Most, Louise M. Nelson, and Melanie D. Jones. 2021. "Soil Organic Carbon in Irrigated Agricultural Systems: A Meta-Analysis." *Global Change Biology* 27 (16): 3898–910. https://doi.org/10.1111/gcb.15680.

Giordano, Mark, Regassa Namara, and Elisabeth Bassini. 2023. "The Impacts of Irrigation: A Review of Published Evidence." World Bank. http://documents.worldbank.org/curated/en/099514502242320808.

Jiang, Yu, Daniela Carrijo, Shan Huang, et al. 2019. "Water Management to Mitigate the Global Warming Potential of Rice Systems: A Global Meta-Analysis." *Field Crops Research* 234: 47–54. https://doi.org/10.1016/j.fcr.2019.02.010.

Ministry of Water, Sanitation and Irrigation, Kenya. 2025. *National Irrigation Sector Investment Plan (NISIP)*. Republic of Kenya.

Oseni, Gbemisola, Kevin McGee, and Andrew Dabalen, 2014. "Can Agricultural Households Farm Their Way Out of Poverty?" Policy Research Working Paper 7093, World Bank. https://openknowledge.worldbank.org/server/api/core/bitstreams/fa906994-3691-5df3-95c0-2d7c96b1d2fc/content.

Qin, Jingxiu, Weili Duan, Stefan Zou, et al. 2024. "Global Energy Use and Carbon Emissions from Irrigated Agriculture." *Nature Communications* 15: 3084. https://doi.org/10.1038/s41467-024-47383-5.

Rodella, Aude-Sophie, Esha Zaveri, and François Bertone. 2023. *The Hidden Wealth of Nations: The Economics of Groundwater in Times of Climate Change*. World Bank.

Sixt, Gregory N., Laurens Klerkx, J. David Aiken, and Timothy S. Griffin. 2019. "Nebraska's Natural Resource District System: Collaborative Approaches to Adaptive Groundwater Quality Governance." Faculty Publications: Agricultural Economics 169. University of Nebraska.

Taheripour, Farzad, and Maksym Chepeliev. Forthcoming. "Assessing Economic Impacts of Changes in Irrigated Area: A Global Analysis Using a Multi-Region Input-Output (MRIO) Framework Based on GTAP 2017 Data Base." Background paper for *Nourish and Flourish: Water Solutions to Feed 10 Billion People on a Livable Planet*. World Bank.

Waalewijn, Pieter, Remi Trier, Jonathan Denison, et al. 2020. *Governance in Irrigation and Drainage: Concepts, Cases, and Action-Oriented Approaches—A Practitioner's Resource*. World Bank.

Wang, Jinxia, Yunyun Zhu, Tianhe Sun, et al. 2020. "Forty Years of Irrigation Development and Reform in China." *Australian Journal of Agricultural and Resource Economics* 64: 126–49.

World Bank. 2021. "Governance in Irrigation and Drainage Appendix 2: Case Descriptions." World Bank.

World Bank. 2025. "Sahel Irrigation Strategy." World Bank. http://hdl.handle.net/10986/43750.

CHAPTER 3

Framework for Regional Strategies

A new framing for agricultural water management: Rooted in local context

Universal transformation of agricultural water management (AWM) will require the three paradigm shifts, but operationalizing them at a local level requires a new water-food nexus framework and defined pathways. In pursuit of water and food security, many countries have developed water resources and expanded irrigated agriculture to grow crops for food, feed, and fiber. National natural water endowments differ widely, as does the intensity of AWM development. Some countries have expanded AWM, particularly irrigation use, to increase agricultural production; others have chosen to rely on trade to secure their agricultural products, thereby conserving their available water.

The new water-food nexus framework is grounded in understanding AWM imbalances or disparities in water resource distribution; that is, some regions underuse local freshwater resources whereas others overuse them. For instance, in Sub-Saharan Africa, local shallow aquifers hold 61 percent of groundwater resources and could irrigate up to 40 million hectares (ha), yet only 7 percent of cultivated land is irrigated (Rodella et al. 2023). Conversely, in arid and semiarid parts of the Middle East, North Africa, and South Asia, overuse is rapidly depleting aquifers, threatening agricultural productivity and rural livelihoods. Achieving a healthy global balance that eliminates both the overuse that degrades water systems and the underinvestment that leaves resources untapped and communities vulnerable is essential to feed the projected 10 billion people on the planet by 2050.

This new framework positions countries along two critical dimensions: freshwater stress levels and food trade status (that is, net calorie importers or exporters). As such, the framework provides a simple yet powerful lens for tailoring AWM policies to different country contexts (figure 3.1). It distills complex, cross-cutting water and food system realities into an intuitive structure that helps policy makers quickly identify priorities, risks,

FIGURE 3.1 Four local contexts defined by water-food nexus framework

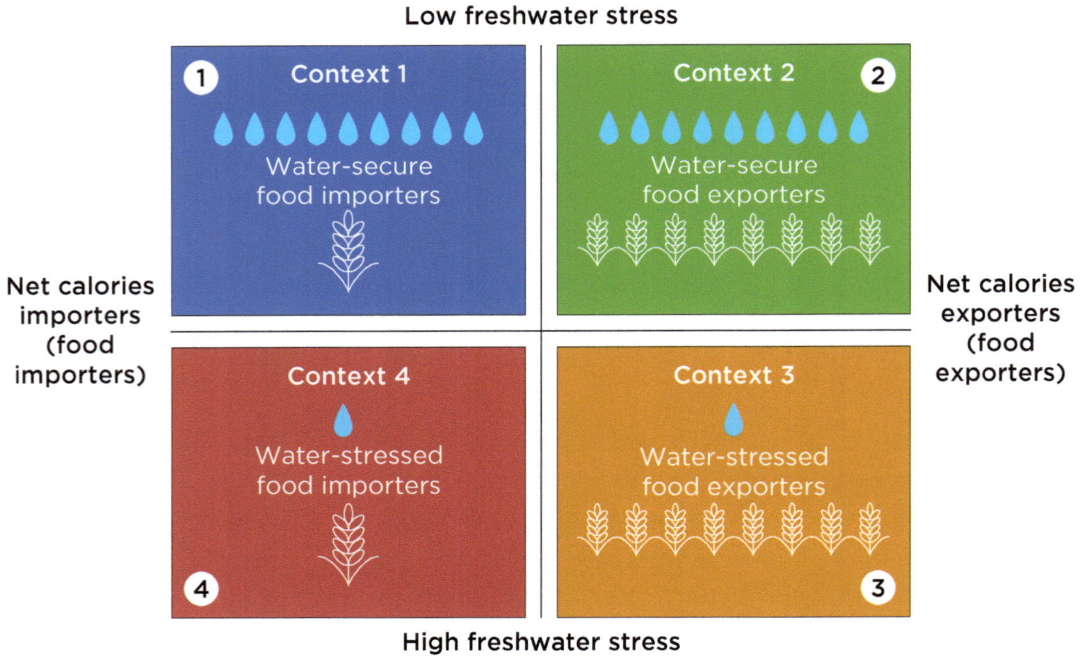

Sources: Food Balance (database), Food and Agriculture Organization of the United Nations (FAO), https://www.fao.org/faostat/en/#data/FBS (food balances); Food Composition Tables, FAO, https://www.fao.org/faostat/en/#data/FBS (to convert food balance items into calories); Aqueduct 4.0 Current and Future Global Maps Data (database), World Resources Institute, https://www.wri.org/data/aqueduct-global-maps-40-data (water data).

and actionable levers tailored to their circumstances (refer to box 3.1). The framework intentionally focuses on food crops; data limitations constrained the analysis of fiber and nonfood export crops (such as coffee and cocoa). Importantly, whereas this framework operates at the national level for policy coherence, most countries contain diverse subnational contexts. AWM strategies must therefore be locally adapted within each country's broader framework and consider market readiness, structural transformation capacity, and value chain maturity considerations.

The water-food nexus framework enables policy makers to identify a starting point for development of a targeted, realistic pathway for change—whether addressing calorie imports, leveraging trade to relieve water resources pressure, or prioritizing AWM investments for maximum economic returns. By identifying where a country is positioned within the water-food nexus framework (map 3.1) and where it is headed, policy makers can design concrete AWM interventions tailored to their specific water-food nexus contexts.

BOX 3.1

The water-food nexus framework methodology explained: Water stress index and calories as a metric for food

The water-food nexus framework organizes countries along two axes (refer to figure 3.1). The vertical axis measures water availability through degrees of blue water stress index as recorded in the Aqueduct 4.0 Current and Future Global Maps Data database of the World Resources Institute, using a 1979–2019 baseline period. Stress levels below 40 percent signal water security; levels above 40 percent indicate water stress.

The horizontal axis captures each country's food trade position based on the Food and Agriculture Organization of the United Nations (FAO) Food Balance data, using a 2010–2019 baseline period. All food commodity imports and exports are represented in terms of a common unit—net calories. This is where an important limitation arises: calorie density varies enormously across food types. According to FAO food composition data, watermelon provides approximately 17 kilocalories per 100 grams (calorie light), while rice provides around 280 kilocalories per 100 grams (calorie dense). A country that exports large volumes of fruits and vegetables that are lighter in calories but imports agricultural commodities that are dense in calories will appear in this framework as a net food importer, because, although the country may export a significant volume of food, more calories are imported than exported. Being a net food importer is not inherently a problem. For some countries, it is a deliberate policy choice—a rational strategy to reduce domestic water use by importing food that would otherwise require significant water resources to produce.

Framework users should note that data quality in the FAO Food Balance sheets varies across countries. The framework also reflects national averages of both water and food status; hence, in-country variation is not captured. Country-level applications could incorporate greater nuance, making the global map a starting point, not a final answer.

This section describes the high-level characteristics of each quadrant of the water-food nexus, and the following section details the pathways.

Context 1: Water-secure food importers. These countries have an unprecedented opportunity for national and global food security. Countries with sufficient freshwater resources but insufficient food production can rapidly expand irrigation and AWM practices in an environmentally responsible manner to reduce dependence on food imports. For countries with reliable access to trade and for countries that

MAP 3.1 World map reflecting four local contexts defined by the water-food nexus framework

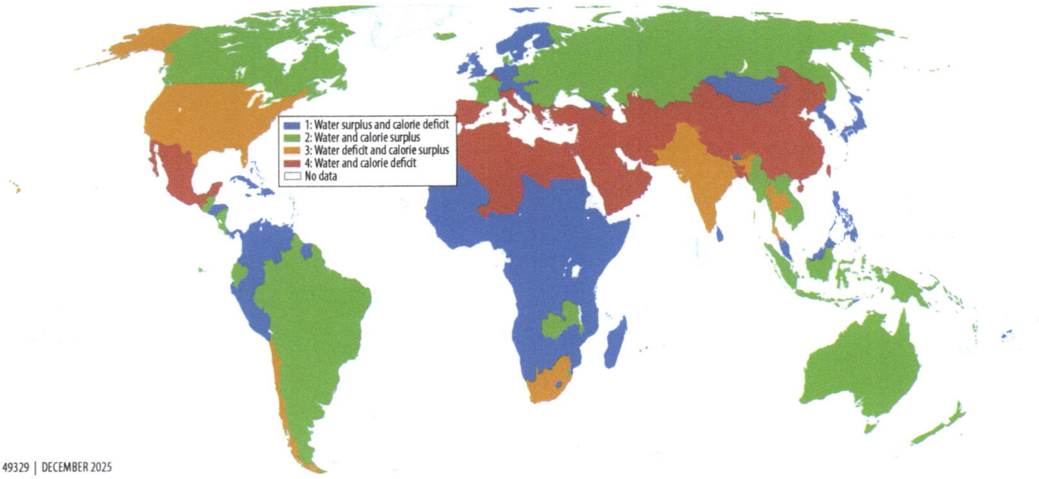

Sources: Food Balance (database), Food and Agriculture Organization of the United Nations (FAO), https://www.fao.org/faostat/en/#data/FBS (food balances); Food Composition Tables, FAO, https://www.fao.org/faostat/en/#data/FBS (to convert food balance items into calories); Aqueduct 4.0 Current and Future Global Maps Data (database), World Resources Institute, https://www.wri.org/data/aqueduct-global-maps-40-data (water data).

lack agroclimatic conditions for specific crops, trade represents an option to access needed calories (while internalizing risks if importing from Contexts 3 and 4).

Context 2: Water-secure food exporters. These countries represent an opportunity for global food security. Countries with sufficient freshwater resources and food production can also pursue an expansion agenda, aiming to strategically and responsibly increase food exports and related economic benefits. The expansion of AWM needs to go hand in hand with the development of local and global value chains for agricultural commodities. From a global perspective, sustainable food production and export could help the global water and food balance by exporting calories to countries that cannot produce needed calories or that choose to rely on trade.

Context 3: Water-stressed food exporters. These countries face an unsustainable contradiction. Nations producing food surpluses while depleting freshwater resources essentially export freshwater they cannot spare, threatening both their ecosystems and future food security. These countries need to prioritize the following: the production of crops that consume less water for equal returns, investment in more efficient and modern irrigation practices, incentives for sustainable farming practices, and the augmentation of irrigation water supply with reused treated wastewater. They should complement these changes with financial, technical, and policy support to buffer the economic and human impacts of any agricultural policy shifts.

Context 4: Water-stressed food importers. These countries confront the challenge of insufficient domestic production coupled with dwindling freshwater resources, creating a precarious dependence on global markets and a need for long-term structural economic change. Transitioning away from farming requires bold investments in alternative livelihoods, infrastructure, and social safety nets to protect vulnerable populations. These countries must focus on economic water productivity (high-value crops with limited water), economic diversification, the augmentation of irrigation freshwater with reused treated wastewater, and imports from secure and diverse sources. Governments need to value and allocate freshwater for maximized economic returns and enforce water use caps and quotas.

Pathways to AWM: Navigation from each water-food nexus context

The water-food nexus framework sets out four typologies in terms of freshwater and calories. By understanding the opportunities, challenges, and choices associated with each of the four contexts of the water-food nexus, policy makers can design pathways to improved AWM. This section describes how these pathways might be practically implemented through a complex landscape of challenges and provides country examples.

Pathway from context 1: Water-secure food importer

Opportunities: Countries can leverage available freshwater resources to increase agricultural production and drive economic transformation. Agricultural production can include the cultivation of high-value crops or strategic agricultural commodities. In this context, food imports will likely continue for most countries and can be managed in terms of exposure and risk. Some countries—even when water secure—may decide to continue to rely primarily on trade because they have reliable trade relationships or revenues to fund trade, place high societal value on environmental outcomes, or lack the appropriate agroclimatic zones for specific crops.

Challenges: When moving from pilots and small-scale successes in AWM to large-scale implementation, countries must stay within local environmental and water thresholds.

Choices: Balance investment decisions between targeting easily accessible areas for quick wins and developing more challenging but potentially higher-impact regions that may require greater up-front investment in infrastructure and capacity building. This choice depends on the source of water (local sources versus infrastructure-dependent options) and funding availability.

Priority actions

- *People.* Focus on income gains for farmers, especially smallholders. Create more and better jobs for workers and farmers. Enhance financial access for smallholders to invest in leveraging freshwater use in AWM.
- *Prosperity.* Invest in freshwater use for increased agricultural production, reducing import dependence or exporting high-value products while importing agricultural commodities. Attract more private sector participation for enhanced performance or financing. Some countries may rely exclusively on trade for food security.
- *Planet.* Accelerate green growth through renewable energy and efficiency, and leverage opportunities from climate grants and carbon financing.

What works and what to scale up: The following examples focus on Burundi, Mali, Senegal, Uganda, and countries of the Sahel region (Burkina Faso, Chad, Mali, Mauritania, Niger, and Senegal). Apart from Niger, these countries are water-secure food importers in the new water-food nexus framework (figure 3.1 and map 3.1).

- *Mali.* The Office du Niger has demonstrated remarkable productivity gains, increasing rice yields from 2 ton/ha to 5–6 ton/ha over 40 years through improved service quality and institutional reforms that give farmers greater control over water and land management. Public funding has been instrumental in developing the infrastructure and irrigation systems, with private sector involvement focused on processing and marketing activities. These improvements have resulted in widespread adoption of high-yield rice varieties. Additionally, joint ventures between international agriculture corporations and the Malian state have occurred within the scheme.
- *Senegal.* The National Company for the Development and Exploitation of the Lands of the Senegal River Delta (Société Nationale

d'Aménagement et d'Exploitation des Terres du Delta du Fleuve Sénégal, or SAED) is a public industrial and commercial establishment operating with a board of directors that provides strategic oversight and accountability. SAED's financial management adheres to national accounting standards, with regular audits and transparent reporting mechanisms. SAED is a major player in the public-private partnership (PPP) space, facilitating the sharing of large-scale irrigation investment costs and responsibilities between public institutions and agricultural producers. As the public partner, SAED is responsible for land management, infrastructure development, and ensuring access to both large-scale and smallholder agricultural entities. It has empowered private producers by fostering large-scale collective schemes and mobilizing private financial resources and expertise, including new irrigation equipment to improve water use efficiency.

- *Uganda.* The Micro-Scale Irrigation Program has successfully mobilized private irrigation through innovative financing. Between 2020 and 2024, the program combined government funding (US$15.4 million) in equipment subsidies (plus additional facilitation costs) with farmer contributions (US$6.3 million). A tiered subsidy structure incentivized the purchase of climate-smart solar pumps, selected by 95 percent of farmers. By 2024, 4,733 participating farmers (21 percent of them women) were cultivating 3,360 ha on an average plot size of 0.71 ha. An economic analysis showed a strong 20-month payback period even without subsidies; however, in such a case, the initially high capital purchase would likely depend on farmers' ability to access affordable financing from commercial banks. The program has prompted systemic changes in the supply chain, with a fourfold increase of suppliers nationally, tested and proven technologies on the ground, and widespread knowledge of irrigation systems and their use. These gains demonstrate the program's financial viability and scalability potential.

Use of data and innovation

- *Burundi.* Spatial analysis using remote sensing and ground-based data helped analyze irrigation needs and priorities for investment across various size categories and project types (rehabilitation of existing systems versus new development). This analysis was used to estimate investment needs across 57,000 ha within peri-urban areas, which in the short to medium term can be developed, given their proximity to existing markets (World Bank 2025).

- *Sahel countries.* The use of machine learning and satellite data for the identification of shallow basement crystalline groundwater potential represents a scalable, data-driven approach that identified nearly 1 million ha of opportunity for irrigation development across five countries, demonstrating how technology can guide targeted, cost-effective irrigation expansion. This innovation has informed the new Sahel Irrigation Strategy approved by all Sahel countries on April 9, 2025.

Pathway from context 2: Water-secure food exporter

Opportunities: Countries can maintain the competitive advantages of agricultural production and export.

Challenges: Countries must ensure that they do not slide toward exceeding local natural resource thresholds.

Choices: Use additional freshwater in AWM within local natural resource thresholds to increase production for export or maintain the current use of freshwater in AWM for agricultural production.

Priority actions

- *People.* Focus on income gains and employment for farmers, especially smallholders.
- *Prosperity.* Improve AWM production with or without increased use of freshwater, and strengthen institutions and markets to maintain or increase revenues from local markets or export.
- *Planet.* Accelerate green growth through renewable energy and efficiency, and leverage opportunities from climate grants and carbon financing.

What works and what to scale up: The following examples focus on France, Peru, and Ukraine. These countries are water-secure food exporters in the new water-food nexus framework (figure 3.1 and map 3.1).

- *France.* Regional Development Companies, such as the Société du Canal de Provence (SCP), demonstrate effective governance through 75-year concessions, diversified revenue streams, and participatory stakeholder boards. SCP provides a scalable model for integrated, financially sustainable water management. As a prominent public irrigation agency, SCP is a public initiative that addresses multiple challenges including land use planning, infrastructure development, risk mitigation, and public awareness to encourage private sector participation. For example, SCP partners with private sector stakeholders, including farmers, to improve water allocation monitoring, reduce energy costs, enhance water reuse, and introduce robotics to improve efficiency, yields, and cost savings. Notably, private Chambers of Agriculture also serve as small shareholders in SCP.

- *Peru.* The Olmos-Tinajones project enabled the irrigation of 43,500 ha (38,000 of newly developed land auctioned to investors and 5,500 for local farmers) through a multistage PPP arrangement. The water transfer stage (about US$247 million) diverted water across the Andes via a dam and a 20-kilometer trans-Andean tunnel, implemented under a PPP that included public support measures such as tax exemptions and a tax stability agreement. The irrigation stage (approximately US$280 million) was privately financed by the irrigation concessionaire and agribusiness investors, with roughly half of the financing raised through local bond issuances supported by a US$50 million partial credit guarantee from the Development Bank of Latin America and the Caribbean. The hydropower component (approximately US$60million to US$80 million) was privately financed under a separate concession. Cost recovery relied primarily on land auctions, which generated at least US$150 million through a land trust mechanism, complemented by a flat water service fee of US$0.07 per cubic meter. By enabling large-scale water transfers, the project facilitated a shift from substance agriculture to high-value, export-oriented production. Although the project experienced delays related to technical challenges, legal disputes, and land acquisition issues, renewed efforts were made in 2024 to resolve these challenges. As with most irrigation PPP arrangements, much of the infrastructure associated with the scheme has been publicly financed. Public contributions included project management, marketing, and selective policy and implementation reforms.

Use of data and innovation

- *France.* SCP achieves high water efficiency through integrated data-driven innovation, with over 85 percent of extracted water invoiced. Real-time sensor networks continuously measure water levels and flows, enabling dynamic regulation from a central Remote-Control Centre to minimize extraction and transportation losses. SCP leverages remote sensing, including manned and unmanned aircraft equipped with hyperspectral cameras for leak detection across its vast network. SCP also develops connected irrigation systems using field sensors and data platforms to provide precise water management for agricultural customers, optimizing conservation and usage.

- *Ukraine.* The use of hydro-informatics and cutting-edge analytics to evaluate the state of Ukraine's irrigated agriculture sector deepened understanding of weather-related risks, particularly water and heat stress risks to agricultural productivity. This was the first time that remote sensing analytics were conducted to analyze the impact of the Russian Federation's invasion of Ukraine on the irrigation sector. The World Bank has conducted extensive remote sensing analysis to assess the impact of Russia's invasion on Ukraine's irrigation sector, primarily through its report "The Future of Irrigation in Ukraine" (Sinha et al. 2023) and a series of Rapid Damage and Needs Assessments. The analysis combined satellite-based remote sensing with census data and earth systems modeling. Cropping intensity—the number of crop cycles observed over a vegetation period—was used as the primary indicator of irrigation performance. Machine learning techniques were applied to map irrigated areas and crop types using satellite imagery, with evapotranspiration and biomass production datasets used to assess crop water demand and irrigation efficiency. Remote sensing analysis of Ukraine's 20 largest irrigation systems showed that in 2022, the amount of land under irrigation fell by approximately 35,500 hectares, or about 13.5 percent, relative to the 2017–2021 average. The Kakhovka irrigation scheme in Kherson registered the largest single drop. The analysis considered risks to irrigation zones, water use for irrigation, biomass production, crop yields, and dams. It provided better understanding of factors such as cropping intensity, crop water productivity, incremental yield, and irrigation efficiency. This analysis supports Ukraine's efforts to "build back better" through well-informed AWM policies and decision-making.

Pathway from context 3: Water-stressed food exporter

Opportunities: Countries can scale up innovations in the agriculture and water sectors to protect and improve ecosystems while maintaining economic development benefits and rural livelihoods.

Challenges: Countries must reverse an unsustainable water security trajectory that threatens long-term benefits for prosperity, people, and planet. Adopting changes in agriculture and water comes with significant costs and challenges in gaining stakeholders' acceptance, particularly with regard to enforcing water allocation. Countries must balance immediate economic benefits from high-productivity agriculture against mounting environmental costs, including groundwater depletion and reduced environmental flows, which will eventually threaten production as well. Policy decisions involve difficult trade-offs between maintaining export revenues and implementing water restrictions that could affect farmer livelihoods. Additionally, shifting to less water-intensive crops or importing food with higher water footprints requires significant market adjustments and may face political resistance.

Choices: Transition to using less freshwater in agriculture or augmenting water sources through nonconventional water sources (desalination and reuse of treated wastewater) and repurposing subsidies to ease the transition from water-intensive agriculture while avoiding devastating rural communities that depend on current production systems.

Priority actions

- *People.* Ensure that reducing freshwater use does not reduce farmer incomes or employment. If it does, include compensation measures and develop employment opportunities in high-value chains.
- *Prosperity.* Increase water productivity with less freshwater. Scale water-saving technologies such as the Chameleon soil moisture sensor and better manage tailwater flows and on-farm irrigation agronomy. Introduce performance-oriented service delivery on public schemes, use nonconventional water with the enforcement of water allocations, and strengthen irrigation and water management institutions (accountability and transparency) for economic transformation.
- *Planet.* Establish and enforce water allocation and water caps based on basin assessments and water trade-offs.

What works and what to scale up: The following examples focus on Chile, India, and Pakistan. These countries are all water-stressed food exporters in the new water-food nexus framework (figure 3.1 and map 3.1).

- *Chile.* The country increased water use efficiency through modernization of irrigation methods, encouraging private farmers to adopt drip and microjet systems. The amount of land using these methods increased from 93,000 ha in 1997 to about 900,000 ha in 2021 (almost 50 percent of total productive land). Government support enabled an increase in land dedicated to high-value crops such as fruit trees and vineyards. Research in Chile found that water availability and socioeconomic characteristics such as land size and access to credit can hinder adoption of irrigation technologies. To overcome these challenges, Chile's National Irrigation Commission created a program supporting small-scale farmers with up to 40 ha (AgriBrasilis 2022). Revisions to the Water Code in 2022 increased attention to resource conservation and socioenvironmental protection in response to public concerns and research showing that irrigation benefits at the farmer level may reduce water returning to rivers and aquifers, with potential negative consequences for other users and ecosystems. Within the Water Code framework, public policies and programs have subsidized irrigation investments to meet agricultural demand and improve efficiency.

- *India.* The Damodar Valley irrigation scheme in West Bengal and Jharkhand incentivizes private sector participation in performance-based contracts with private sector irrigation service providers. This new approach aims to ensure sustained operational efficiency, enhance quality of service delivery, and reduce dependency on ground water. Still under implementation, a key innovation of this approach has been the engagement of nine irrigation service providers for operation and maintenance of sections of the irrigation scheme. Farmers have seen enhanced irrigation services, and the area supplied with canal irrigation has increased during two (Boro and Rabi) of three total harvest seasons.
- *Pakistan.* In 1984, the provincial agricultural department in Punjab province initiated laser land leveling, a high-precision farming technique that uses lasers to create a perfectly flat or uniformly sloped field. To accelerate leveling, the Punjab government shifted service delivery to the private sector in 2006, supporting this transition with a 50 percent matching grant program to enable private providers to acquire equipment. Consequently, annual laser land leveling capacity in the province increased from 14,000 ha to 300,000 ha per year, leading to the complete transfer of remaining government units to the private sector by 2016 and resulting in operation of more than 16,000 private units across the province. Over time, the private sector has played a primary role in implementing laser land leveling in Pakistan, particularly in Punjab. Services include equipment, knowledge transfer, and technology that can be purchased directly from manufacturers and suppliers or rented by small-scale farmers who may not be able to afford the initial investment in this technology.

Use of data and innovation

- *India.* The World Bank–supported National Hydrology Project launched the modernization of states' water resources management departments through advanced hydro-informatics, real-time flood forecasting, and integrated basin-scale planning. The project includes components such as the establishment of the National Water Informatics Center, which standardizes hydrological data across agencies and makes the data publicly accessible, and the deployment of decision-support systems leveraging artificial intelligence for reservoir operations and drought prediction. Challenges remain regarding the systematic use of data for decisions, although the availability now of prerequisites represents a major achievement of the project.

Pathway from context 4: Water-stressed food importer

Opportunities: Countries can pioneer integrated solutions combining innovations in the agriculture and water sectors, livelihood diversification, and demand-side management in food systems (including diets, waste, and losses) to improve ecosystem health and reduce the risk of trade-related shocks, while maintaining economic development benefits and supporting rural livelihoods.

Challenges: Countries must coordinate investments and policy innovations to create new pathways for rural development, ecosystem protection, and food security in a resource-constrained context. A key challenge lies in simultaneously reducing water consumption and maintaining agricultural livelihoods in water-dependent communities. Countries face difficult choices between immediate food production needs and long-term water sustainability, often under pressure from growing populations and climate shocks. Dependence on food imports creates vulnerability to global market volatility and geopolitical tensions, and domestic water restrictions may exacerbate this dependence. The need to shift farmer livelihoods to nonagricultural sectors or high-value, low-water-use crops compounds this challenge, because the shift requires substantial investment in education, infrastructure, and market development, which often strains already limited public resources. Water-scarce settings typically have fewer options to incentivize and enforce water limits, and existing options are more costly.

Choices: Transition to using less freshwater in agriculture or augmenting water availability through nonconventional sources, such as desalination, treated wastewater reuse, or interbasin transfers.

Repurposing subsidies could ease the transition away from water-intensive agriculture, thus minimizing negative impacts on rural communities dependent on current production systems. Promote shifts in diets away from water-intensive foods, and support reduction of food waste and losses. Balance domestic production with strategic objectives, such as grain self-reliance or safeguarding high-value crops, to enhance resilience to trade shocks.

Priority actions

- *People.* Ensure that reduced freshwater use does not negatively affect farmer incomes and employment; where impacts occur, introduce compensation measures and promote alternative livelihoods.
- *Prosperity.* Increase productivity using less freshwater, augment water supply through nonconventional sources, repurpose subsidies to support the transition to high-value crops and less water-intensive crops, and strengthen resilience to trade shocks. Enhance financial access to water-saving technologies with enforcement of water quotas.
- *Planet.* Establish and enforce water allocation based on basin assessments and water trade-offs, promote less water-intensive diets, and reduce food waste and losses.

What works and what to scale up: The following examples focus on Jordan, Morocco, Spain, and Uzbekistan. These countries are all water-stressed food importers in the new water-food nexus framework (figure 3.1 and map 3.1).

- *Jordan.* The As-Samra Wastewater Treatment Plant shows how treated wastewater reuse can supplement freshwater for irrigation. Developed through a 25-year build-operate-transfer arrangement, and with a political risk guarantee provided by the Multilateral Investment Guarantee Agency, the plant was designed, built, and operated under a PPP arrangement. The primary entity, the Samra Wastewater Treatment Plant Company Limited, has responsibility for securing private financing, expanding the plant, and operating and maintaining it according to a long-term concession agreement with the Jordanian government. The concession runs until 2037, after which the plant will revert to government ownership.
- *Morocco.* A risk-sharing facility de-risks lending for on-farm solar and irrigation investments, offering a replicable model for enhancing financial access to water-saving technologies while promoting renewable energy in agriculture. The facility's partners include the Banque Centrale Populaire, based in Morocco; the Compagnie

Marocaine de Goutte-à-Goutte et de Pompage, Morocco's largest manufacturer and distributor of irrigation systems and water infrastructure; and the International Finance Corporation. The Banque Centrale Populaire provides loans primarily to farmers and micro and small enterprises in the distribution company's network to support expansion of microirrigation and solar energy. The new systems replace fuel-run systems, thus reducing greenhouse gas emissions. The risk-sharing mechanism mitigates lending risk and encourages private sector participation in sustainable agriculture in Morocco. Nearly 30,000 farmers are expected to benefit from the new equipment by 2027.

- *Spain.* Between 1996 and 2017, 50 percent of the country's 3.7 million ha of irrigated land was modernized to improve service efficiency, agricultural competitiveness, and environmental sustainability. Interventions included farmers' adoption of localized or sprinkler irrigation systems and water user associations' placement of solar panels on floating structures within regulation reservoirs, improving the panels' performance, reducing evaporation, and limiting algae growth. These achievements were accomplished through a PPP with the State Society for Agricultural Infrastructure (SEIASA), which advanced 50 percent of the total project execution cost; the target community paid the balance. The irrigation districts provide the land needed for the projects, and SEIASA will own the systems until private sector entities, including farmers, can repay the loans associated with the irrigation investments. Of the €2 billion worth of investment that SEIASA has carried out over the last 20 years, approximately 48 percent of the funding came from SEIASA's own contributions, 29 percent from irrigators, and 23 percent from European Union funds and regional government funds.

Use of data and innovation

- *Uzbekistan.* The National Irrigation and Energy Efficiency Improvement Project aims to strengthen digital technologies and use of remote sensing to improve irrigation service delivery. The project supports the integration of remote sensing tools to monitor irrigation performance, water consumption, and productivity, and to inform water accounting and decision-making processes. These efforts are intended to enhance the efficiency, transparency, and climate resilience of irrigation management. Early-stage investments in digital solutions are critical to enable data-driven planning, improve equity in water distribution, and support long-term transitions to more financially and environmentally sustainable service delivery.

References

AgriBrasilis. 2022. "Chile's Agriculture Depends on Irrigation Systems." *AgriBrasilis,* July 20, 2022. https://agribrasilis.com/2022/07/20/chiles-agriculture-depends-on-irrigation-systems/.

Rodella, Aude-Sophie, Esha Zaveri, and François Bertone. 2023. *The Hidden Wealth of Nations: The Economics of Groundwater in Times of Climate Change.* World Bank.

Sinha, Ranu, Poolad Karimi, Lorenzo Rosa, and Silvan Ragettli. 2023. "The Future of Irrigation in Ukraine." World Bank. https://documents1.worldbank.org/curated/en/099062524074575405/pdf/P180198-667649a2-2436-436c-b4fc-484bbd031258.pdf.

World Bank. 2025. "Rooted in Opportunity: Shaping the Future of Burundi's Agri-Food Economy—A Roadmap for a Prosperous Agri-Food System in Burundi." World Bank.

CHAPTER 4

Funding and Policy Development

Financing agricultural water management

As countries consider how to implement transformational change in agricultural water management (AWM), they face a key concern: funding requirements. Countries will need to rethink current approaches that are rooted in fragmented, investment-centric models and overdependence on public investment funding, and that often lack performance accountability or emphasis on long-term sustainability. Traditional approaches—reliant on ad hoc subsidies or siloed grants—fail to incentivize scalable outcomes, and high transaction costs and repeated reinvention of solutions exacerbate challenges for smallholders and underserved markets.

Expanding and modernizing irrigation to meet the global demand for food will require an estimated public capital investment of US$600 billion to US$1.8 trillion—or US$24 billion to US$70 billion annually from 2025 to 2050 (in 2024 prices). This report presents an estimated range of global capital investments required for both the sustainable conversion from rainfed to irrigated agriculture and the modernization of unsustainable irrigation areas, using the literature to estimate the area for conversion and modernization (Rosa et al. 2020). A portfolio review of World Bank–funded irrigation projects was conducted to extract the unit cost of various irrigation interventions given the lack of a current global dataset on unit costs for AWM, including irrigation, over the past 15 years. The estimated cost is based on the area and the unit cost (Li et al., forthcoming).

More effective use of existing budgets and repurposing of existing agricultural support and subsidies (Damania et al. 2023) may be adequate to meet priority needs in many countries, thus averting an immediate need to mobilize additional funding for AWM. In other countries, however, underinvestment in the sector as a whole remains the predominant challenge. Globally, proxied by data from 90 countries, agricultural public expenditure reached almost US$490 billion in 2023.[1] Three-fourths of this expenditure was used for direct farm payments, with most of the amount used to finance output or input subsidies and payments linked to production of specific commodities. This support distorts production and consumption decisions, causing harmful environmental impacts (Damania et al. 2023). In 2023, governments spent US$106 billion on general agricultural support services, such as research and extension services, rural roads, and other market infrastructure, including US$27 billion for irrigation. Shifting a portion of agricultural public expenditures to irrigation, along with other actions discussed in this report, would reduce the financing gap for irrigation expansion.

For on-farm investments, although highly profitable, many farmers, particularly subsistence farmers, are not deemed creditworthy or lack access to the appropriate pools of credit. Government plays a critical role in connecting smallholder farmers—84 percent of the world's 570 million farmers—with financing options. Government support in creating an enabling environment for smallholders and the private sector (suppliers, value chain actors, and financial institutions) to access finance, act collectively, innovate, and interact can have a major impact on local food production, the economy, and surrounding landscapes. Blended funding can help farmers to access agroecological innovations such as rainwater harvesting, agroforestry, and diversified cropping systems; increase access to extension services; and establish local and international market links.

Off-farm infrastructure investments need to be paired with sustainable revenue streams for operation and maintenance (O&M). When irrigation services are mostly funded through the public sector, funding is typically unreliable and performance is low because of a lack of accountability to water users—typically considered beneficiaries rather than consumers. In addition, the sector has tended to focus on new greenfield AWM infrastructure while ignoring service quality, O&M costs, the technical and financial efficiency of the systems, and the administrative, accountability, and regulatory regimes that oversee this infrastructure. To address financial sustainability and attract investment, irrigation must shift from a "design, build, neglect, and rebuild" cycle to a model of sustained investment and O&M financed, at least in part, by users.

O&M costs for irrigation schemes vary by system type (drip, center pivot, or surface), input prices, and external factors. Key O&M inputs include fuel and lubrication, labor, maintenance of fields and reservoirs, and parts; external factors include the quality of source water and the age and condition of the system. Given the variability, no global benchmark exists for irrigation O&M costs but available data (primarily from India and the United States) indicate that O&M budgets in the range of 3 to 8 percent of initial costs are appropriate.[2]

Evidence shows that farmers readily pay for dependable services, at levels that can cover O&M costs. Willingness to pay is apparent in informal groundwater markets, in which farmers pay high prices for water that is accessible on demand. Willingness to pay accompanies improved service contracts, short-route accountability to users, and data transparency and contestability (World Bank 2021). New tools like digital payment, such as Pakistan's e-abiana and e-voucher systems, offer a way to streamline and facilitate farmer payments (World Bank 2025). However, the sustained functioning of irrigation systems will require the institutionalization of regulatory approaches that incorporate coverage of efficient, real O&M costs.

Governments have a critical role to play in regulation, ensuring transparent methodologies for cost recovery including O&M and asset replacement costs. Consistent policy and legal and regulatory frameworks create an enabling environment for a customer service orientation, strengthen accountability, and support rate-setting to recover efficient costs. India's Maharashtra Water Resources Regulatory Authority, for example, sets bulk water tariffs for irrigation and domestic and industrial sectors, aiming to ensure the embedding of O&M costs in tariff structures. Effective regulation requires that objectives, form, and function align with political economy realities and the institutional framework of the country (Mumssen et al. 2018).

Once stable revenue streams are established, a range of financial instruments becomes feasible. These instruments include microfinance, vendor finance, commercial finance, bonds, guarantees, and public-private partnerships (PPPs). Financing solutions must be tailored to the users and calibrated to the size and scope of investment needed, the creditworthiness of the borrower, the legal and regulatory framework of the jurisdiction, liquidity levels, interest rates, terms, and currencies. Public and private resources can be blended to meet AWM financing needs; for instance, companies can leverage philanthropic capital to de-risk commercial investment, employ results-based financing that pays for demonstrated outcomes rather than inputs, or use carbon credits to subsidize, for example, solar irrigation pumps.

Given the high up-front capital costs, scaling AWM will require long-term financing partnerships. Successful AWM initiatives depend on strategic collaboration among governments, donors, financial service providers, equipment suppliers, value chain alliances, and farmers to align resources, expertise, governance, and accountability needed for sector growth. PPPs demonstrate how risk-sharing and long-term concessions between public and private partners can attract private capital and deliver reliable agricultural water services. In Uganda, as of 2024, the Micro-Scale Irrigation Program demonstrated a successful financing partnership between government and farmers, with government mobilizing US$15.4 million and farmers mobilizing US$6.3 million (29 percent private capital mobilization) toward the cost of 4,733 irrigation installations (Denison and Christen 2025).

A structured approach to AWM policy development

In the complex landscape of AWM—in which diverse policy domains, competing objectives, multiple stakeholders, and varying time frames intersect—policy makers must pursue a practical and coherent policy process to align visions and balance trade-offs.

Process

The water-food nexus framework and associated pathways serve as the starting point for providing AWM policy makers with a foundation for weighing trade-offs across the three strategic policy objectives: people, prosperity, and planet. Building from this foundation, policy makers follow an iterative process, summarized in the following steps:

1. Diagnose the baseline through rigorous assessments.
2. Set coordination mechanisms across sectors.
3. Provide direction using data, priorities, incentives, and feedback loops.
4. Integrate performance-based service delivery.
5. Diversify and reconcile financing sources (public and private).

An example from China, where systemic policy adjustments have aligned with evidence-based decision-making and cross-sectoral coordination, demonstrates the structured approach to AWM policy in action. Over the past few decades, China has shifted from infrastructure-driven development to treating water as a binding constraint on growth. Specifically, the AWM transition involved a move from traditional water-intensive irrigation practices to more sustainable and productive agricultural water use practices. This transformation occurred over the following phases:

1. Developing and upgrading infrastructure to increase irrigation water access

2. Modernizing irrigation management and implementing pricing reforms
3. Integrating irrigation with agricultural practices to increase resource productivity
4. Transforming AWM through data, basin planning, and market-based tools—such as evapotranspiration (ET) management and enforcement of ET quotas.

China has advanced water strategies from local to national levels, notably through the Three Red Lines policy—which sets binding limits on use, efficiency, and quality—and by focusing on climate-smart and sustainable irrigation services, supported by ecological restoration and economic instruments. For example, the seasonal land fallowing policy rewards farmers for reducing water-intensive crops.

China's evolving water governance offers valuable lessons for areas facing chronic water scarcity and development pressures. A comprehensive planning system—which integrates development, spatial, and water management objectives—backs the reforms. An adaptive "start small, learn fast, transform quickly" approach enabled the national scaling of successful pilots. In each phase, China introduced new governance dimensions, reflecting lessons learned and an expanded understanding of water's role in ecological, economic, and social systems.

Although efforts continue to enhance AWM, China's transformation has already delivered tangible outcomes. The country transitioned from a net virtual water exporter to a net virtual water importer while ensuring grain security. It has maintained over 95 percent self-sufficiency in major grains and built reserves sufficient to meet domestic demand for a year or more. China's comprehensive AWM strategy balances trade-offs between prosperity, people, and planet, enabling it to feed 18 percent of the world's population with just 9 percent of the world's cultivated land and 6 percent of the world's water resources.

Policies

Table 4.1 highlights high-priority policies and related actions, applicable to each context, that governments can undertake or drive. Each context is averaged, because some differences occur between countries within a context and within each country, based on sublocations. Refer to the technical volume for this report for additional guidance and details (Li et al., forthcoming).

TABLE 4.1 Transforming AWM to feed 10 billion people on a livable planet by 2050

Context 1: Water-secure food importers	Context 2: Water-secure food exporters	Context 3: Water-stressed food exporters	Context 4: Water-stressed food importers
These countries may choose to use trade to access needed calories when they have reliable access to trade or lack the agroclimatic conditions for specific crops. Context 1 also includes agriculture-dependent economies that want to reduce reliance on food imports. This table is oriented to those countries that choose to rely on their water resources to expand irrigation and reduce reliance on imports.	*These countries can pursue an expansion agenda, aiming to strategically and responsibly increase food exports and related economic benefits. Exports from these countries can help the global water and food balance by exporting calories to countries that cannot produce needed calories, or that choose to rely on trade.*	*These countries export food while further depleting water resources, essentially exporting water they cannot spare and threatening both their ecosystems and future food security. These countries can transition to AWM/irrigation systems that consume less water or transition from a reliance on agriculture.*	*Countries with insufficient domestic food production and stressed water resources face a precarious dependence on global markets. These countries need to reduce water consumption but may remain net food importers for some time given the lack of freshwater.*
Countries in Contexts 1 and 2 are typically characterized by the availability of blue water and have multiple economic opportunities—as importers or exporters—based on freshwater availability. Investment in weather and hydrological information and services will be essential to support the conversion from rainfed agriculture to irrigation or to increase agricultural production from rainfed agriculture in the absence of available local blue water.		Countries in Contexts 3 and 4 experience a level of blue water stress that necessitates reducing blue water consumption in irrigation or an economic structural transformation, particularly in Context 4. Countries that transition away from farming require bold investment in alternative livelihoods, infrastructure, and social safety nets to protect populations adversely affected by change. Governments must appropriately value and allocate water for maximized economic returns and enforce water use caps. Investment in weather and hydrological information and services will be essential to support the agricultural sector in reducing irrigation water consumption or to support transitions away from agriculture.	

(continued)

TABLE 4.1 Transforming AWM to feed 10 billion people on a livable planet by 2050 *(continued)*

Policy recommendation 1: Invest in AWM to increase productivity and reduce hydrological risks.			
Application to Context 1: **Water-secure food importers**	**Application to Context 2:** **Water-secure food exporters**	**Application to Context 3:** **Water-stressed food exporters**	**Application to Context 4:** **Water-stressed food importers**
Increase agricultural productivity through AWM investment. Key actions: • Improve crop water use in rainfed agriculture through better seeds and agricultural practices in areas with no local blue water availability or where the use of blue water is not economically justified. • Supplement rainfall deficits with blue water sources to boost agricultural yields, without exceeding local blue water limits.	Maintain or increase agricultural productivity through AWM investment. Key actions: • Improve crop water use in rainfed agriculture through better seeds and agricultural practices in areas with no local blue water availability or where the use of blue water is not economically justified. • In localized areas with green water scarcity, use blue water to maintain or boost agricultural yield, without exceeding local blue water limits.	Focus on AWM interventions that minimize water consumption. Key actions: • Reduce blue water use through adoption of drought-resistant crop varieties, soil moisture conservation techniques, supplemental irrigation technologies, and fully controlled irrigation. • Augment irrigation water supply with reused treated wastewater or desalinated water. • Increase the availability of data, information, and technologies to support the reduction of water consumption in agriculture.	Focus on economic water productivity—high-value crops produced with limited water. Key actions: • Augment irrigation water sources with treated wastewater or desalinated water. • Develop, monitor, and enforce water consumption caps.

(continued)

TABLE 4.1 Transforming AWM to feed 10 billion people on a livable planet by 2050 *(continued)*

Policy recommendation 2: Enhance smallholder farmers' access to finance, and mitigate risks to catalyze investment.			
Application to Context 1: **Water-secure food importers**	**Application to Context 2:** **Water-secure food exporters**	**Application to Context 3:** **Water-stressed food exporters**	**Application to Context 4:** **Water-stressed food importers**
Enhance smallholder farmers' ability to adopt AWM solutions, including access to finance for private irrigation. Key actions: • Develop financial products (affordable credit, microloans, targeted subsidies, and grants) that enable vulnerable farmers or specific value chains to expand production through irrigation investment. • De-risk private investment for conversion from rainfed agriculture to irrigation through blended finance mechanisms such as public guarantees, first-loss capital provisions, and viability gap financing. • Support the conversion from rainfed agriculture to irrigation through adoption of certification standards and tax policies that protect farmers from overpricing or poor-quality products and lower the cost of essential equipment. • Develop strategic partnerships among local cooperatives and digital platforms to extend AWM services to remote farmers.	Within a well-functioning AWM sector, identify and address opportunities for improved performance. Key action: • Develop tailored financial products to reach underserved smallholders within an otherwise well-functioning sector.	Enhance smallholder farmers' ability to adopt and scale approaches and technologies that reduce water consumption in irrigation. Key actions: • Develop tailored financial products for vulnerable farmers to support the transition to crops and practices that consume less water. • Support farmers' access to finance to scale technologies and practices that reduce water consumption.	Enhance smallholder farmers' ability to reduce water consumption in irrigation and potentially transition away from agriculture. Key actions: • Develop tailored financial products for vulnerable farmers to support the transition to crops and practices that consume less water. • Develop partnerships, platforms, and financial products that support farmers to transition out of agriculture to other employment.

(continued)

TABLE 4.1 Transforming AWM to feed 10 billion people on a livable planet by 2050 *(continued)*

Policy recommendation 3: Strengthen AWM institutions and markets to drive economic transformation.			
Application to Context 1: **Water-secure food importers** Strengthen AWM institutions and markets, including rural economies, to sustain economic growth. Key actions: • Build or strengthen AWM institutions with clear mandates and adequate resources and capacity. • Reduce market risks by strengthening market links (buyers, suppliers, and financial institutions).	**Application to Context 2:** **Water-secure food exporters** Strengthen AWM institutions and markets, including rural economies, to sustain economic growth. Key actions: • Strengthen AWM institutions with clear mandates and adequate resources and capacity. • Strengthen trade strategies for accessing premium markets. • Pair AWM investments with complementary transportation and cold storage to maximize trade potential.	**Application to Context 3:** **Water-stressed food exporters** Strengthen AWM institutions and markets, including rural economies, to drive economic growth within sustainable water constraints. Key actions: • Develop markets for crops produced with less water. • Strengthen trade strategies for accessing premium markets. • Introduce water conservation regulations and include an accountability matrix for AWM institutions.	**Application to Context 4:** **Water-stressed food importers** Strengthen AWM institutions and markets to drive economic growth, including rural economic transformation, within sustainable water constraints. Key actions: • Develop markets for crops produced with less water. • Diversify imports from countries to reduce risk of insufficient food supply. • Introduce water conservation regulations and include an accountability matrix for AWM institutions.
Policy recommendation 4: Enhance food security through trade and strategic grain reserves.			
Application to Context 1: **Water-secure food importers** Expand agricultural production, and secure import sources and terms. Key actions: • Diversify trading partners and refine or optimize contracts and agreements to reduce import risks. • Balance import substitution and export earnings through targeted value chain support.	**Application to Context 2:** **Water-secure food exporters** Maintain or expand food production for trade benefits (exports). Key actions: • Secure or expand trading partners in support of export strategies. • Establish strategic grain reserves to buffer against climate and market shocks.	**Application to Context 3:** **Water-stressed food exporters** Adapt agricultural production and export strategies to freshwater availability. Key action: • Reexamine and adapt export strategies in the context of water stress, reflecting necessary shifts in crop types or production levels.	**Application to Context 4:** **Water-stressed food importers** Adapt agricultural production to water availability while advancing security of food imports. Key actions: • Diversify trading partners and refine or optimize contracts and agreements to reduce import risks. • Establish strategic grain reserves to buffer against climate and market shocks.

(continued)

TABLE 4.1 Transforming AWM to feed 10 billion people on a livable planet by 2050 *(continued)*

Policy recommendation 5: Focus AWM solutions on employment and income gains for all farmers—with emphasis on smallholders.			
Application to Context 1: **Water-secure food importers**	**Application to Context 2:** **Water-secure food exporters**	**Application to Context 3:** **Water-stressed food exporters**	**Application to Context 4:** **Water-stressed food importers**
Increase food production to maximize employment opportunities and income gains for farmers. Key actions: • Prioritize profitable, market-aligned production systems that provide stable, year-round income for farmers. • Prioritize interventions with smallholder-focused, labor-intensive, and high-employment value chains to maximize employment and better jobs.	Maintain or increase food production to maximize employment opportunities and income gains for farmers. Key action: • Prioritize profitable, market-aligned production systems to maximize employment and better jobs.	Realign AWM interventions toward consideration of water availability while seeking to protect agricultural jobs or support employment transitions. Key actions: • Prioritize profitable, market-aligned production systems within freshwater availability limits with the aim of protecting jobs to the extent possible. • Balance efforts to reduce agricultural consumption of freshwater with social safety nets to support farmers who lose jobs or experience reduction in income.	Realign AWM interventions toward consideration of water availability while seeking to protect agricultural jobs or support employment transitions. Key actions: • Prioritize profitable, market-aligned production systems within water availability limits with the aim of protecting jobs to the extent possible. • Balance reduction of agricultural consumption of freshwater with social safety nets and alternative employment options (outside of agriculture) to support farmers who lose jobs or experience reduction in income.

(continued)

TABLE 4.1 Transforming AWM to feed 10 billion people on a livable planet by 2050 *(continued)*

<table>
<tr><th colspan="4">Policy recommendation 6: Establish and enforce water allocation and land zoning based on basin level assessments.</th></tr>
<tr><td>Application to Context 1:
Water-secure food importers
Increase food production while adhering to local water and land availability thresholds.
Key action:
• Convert rainfed agriculture to irrigation where freshwater is locally available and within local availability thresholds.</td><td>Application to Context 2:
Water-secure food exporters
Maintain or increase food production while adhering to local water and land availability thresholds.
Key actions:
• Convert rainfed agriculture to irrigation in areas with green water scarcity where freshwater is locally available and within local availability thresholds.
• Determine green and blue water availability for future production, and adopt strategies to convert rainfed agriculture to irrigation or to phase out agriculture where green water is projected to become scarce.</td><td>Application to Context 3:
Water-stressed food exporters
Establish and enforce water resource allocation and land zoning systems that safeguard stressed water resources.
Key action:
• Establish and enforce data-based water allocation and land zoning systems to reduce the use of freshwater in irrigation.</td><td>Application to Context 4:
Water-stressed food importers
Establish and enforce water resource allocation and land zoning systems that safeguard stressed water resources.
Key actions:
• Urgently establish and enforce data-based water allocation and land zoning systems to reduce the use of freshwater in irrigation.
• Reallocate water to higher-value crops or higher-value uses outside of agriculture.</td></tr>
</table>

(continued)

TABLE 4.1 Transforming AWM to feed 10 billion people on a livable planet by 2050 *(continued)*

Policy recommendation 7: Accelerate green growth using renewable energy, and reduce energy use in irrigation.			
Application to Context 1: **Water-secure food importers**	**Application to Context 2:** **Water-secure food exporters**	**Application to Context 3:** **Water-stressed food exporters**	**Application to Context 4:** **Water-stressed food importers**
Facilitate transition to renewable energy, attracting climate grants/carbon markets to support the financing of increased food production. Key action: • Incentivize farmers, especially smallholders, to adopt renewable energy in irrigation, including by facilitating access to climate grants and carbon markets, thereby supporting investment in converting rainfed agriculture to irrigation.	Facilitate transition to renewable energy, attracting climate grants/carbon markets to support the financing of maintaining or increasing food production. Key actions: • Incentivize farmers, especially smallholders, to adopt renewable energy in irrigation in areas with green water scarcity, including by facilitating access to climate grants and carbon markets, thereby supporting investment in converting rainfed agriculture to irrigation in these areas. • Incentivize modernization of aged irrigation schemes for greater water and energy efficiency and cost savings in operation and maintenance.	Facilitate transition to renewable energy to enhance water productivity and reduce environmental impact. Key action: • Incentivize farmers, especially smallholders, to adopt renewable energy in irrigation, including by facilitating access to climate grants and carbon markets, thereby supporting investment in irrigation systems that achieve cost savings and reduce irrigation water consumption at the basin level.	Facilitate transition to renewable energy to enhance water productivity and reduce environmental impact. Key action: • Incentivize farmers, especially smallholders, to adopt renewable energy in irrigation, including by facilitating access to climate grants and carbon markets, thereby supporting investment in irrigation systems that achieve cost savings and reduce irrigation water consumption at the basin level.

(continued)

TABLE 4.1 Transforming AWM to feed 10 billion people on a livable planet by 2050 *(continued)*

Policy recommendation 8: Promote water-smart food system through producer incentives and consumer awareness of the water footprint of various diets.			
Application to Context 1: **Water-secure food importers**	**Application to Context 2:** **Water-secure food exporters**	**Application to Context 3:** **Water-stressed food exporters**	**Application to Context 4:** **Water-stressed food importers**
Improve the nutritional value of food production. Key action: • Provide incentives for farmers to grow nutrient-dense crops within local freshwater availability thresholds.	Ensure that food production and export strategies stay within resource availability thresholds. Key action: • Provide incentives to farmers to grow nutrient-dense crops, including for demand-driven exports within local freshwater availability thresholds.	Lower the water footprint of food systems by promoting plant-based diets. Key action: • Provide incentives to farmers to grow nutrient-dense crops with low water requirements, and repurpose subsidies away from water-intensive crops.	Lower the water footprint of food systems by promoting plant-based diets. Key action: • Prioritize incentives to farmers to grow nutrient-dense crops with low water requirements, and repurpose subsidies away from water-intensive crops.

Source: World Bank.
Note: AWM = agricultural water management.

Notes

1. Based on 2025 data from Global Alliance for Food Security's Global Food and Nutrition Security Dashboard, https://www.gafs.info/home/. Total agricultural support in 2023 was estimated at US$663 billion, which included US$175 billion of market price support (that is, a transfer from consumers to farmers as a result of import protection). Because this support is not readily available for governments to repurpose for irrigation infrastructure, for example, this report uses the estimate of public expenditures only.
2. In the absence of comprehensive global studies of irrigation O&M, estimate is derived from Ağızan and Bayramoğlu (2021); AgriStuff (2025); Sagardoy, Bottrall, and Uittenbogaard (1982); Skutsch (1998); Skutsch and Evans (1999); UT Extension (2002); and Irrigation Toolbox, US Department of Agriculture, https://irrigationtoolbox.com/WebPages/extensiondocs.html.

References

Ağızan, Süheyla, and Zeki Bayramoğlu. 2021. "Comparative Investment Analysis of Agricultural Irrigation Systems." *Tekirdağ Ziraat Fakültesi Dergisi* 18 (2): 222–33. https://doi.org/10.33462/jotaf.745548.

AgriStuff. 2025. "Agricultural Irrigation Systems Compared: Drip, Pivot, Furrow—Costs & Efficiency." *AgriStuff*, October 18, 2025. https://agristuff.com/farming/agricultural-irrigation-systems-compared-drip-pivotfurrow-costs-efficiency/.

Damania, Richard, Esteban Balseca, Charlotte de Fontaubert, et al. 2023. *Detox Development: Repurposing Environmentally Harmful Subsidies*. World Bank. http://hdl.handle.net/10986/39423.

Denison, Jonathan, and Evan Christen. 2025. "Learning Note on the Design and Rollout of the Uganda Micro-Scale Irrigation Program: Enabling Farmer-Led Irrigation Development by Closing the Affordability and Knowledge Gap." World Bank.

Li, Ruyi, Francois Onimus, Poolad Karimi, et al. Forthcoming. "Portfolio Review of World Bank Irrigation Project, Unit Costs and Total Financing Need." Background note for *Nourish and Flourish: Water Solutions to Feed 10 Billion People on a Livable Planet*. World Bank.

Mumssen, Yogita, Gustavo Saltiel, and Bill Kingdom. 2018. "Aligning Institutions and Incentives for Sustainable Water Supply and Sanitation Services." World Bank. https://openknowledge.worldbank.org/handle/10986/29795.

Rosa, Lorenzo, Davide Danilo Chiarelli, Matteo Sangiorgio, et al. 2020. "Potential for Sustainable Irrigation Expansion in a 3°C Warmer Climate." *Proceedings of the National Academy of Sciences* 117 (47): 29526–34.

Sagardoy, J. A., Anthony Bottrall, and G. O. Uittenbogaard. 1982. *Organization, Operation and Maintenance of Irrigation Schemes*. Food and Agriculture Organization of the United Nations.

Skutsch, J. C. 1998. "Maintaining the Value of Irrigation and Drainage Projects." TDR Project R 6650, Report OD/TN 90, HR Wallingford. https://assets.publishing.service.gov.uk/media/57a08da4ed915d3cfd001b38/R6650-odtn90.pdf.

Skutsch, John, and Darren Evans. 1999. *Realizing the Value of Irrigation System Maintenance*. Issues Paper No. 2. International Programme for Technology and Research in Irrigation and Drainage, Food and Agricultural Organization of the United Nations.

Talbi, Amal, Pieter Waalewijn, Poolad Karimi, et al. 2026. *Nourish and Flourish: Water Solutions to Feed 10 Billion People on a Livable Planet. Technical Background Volume*. World Bank.

UT Extension (University of Tennessee Agricultural Extension Service). 2002. "PB1721-Irrigation Cost Analysis Handbook." PB1721-1M-10/02 E12-4315-00-003-03, UT Extension. https://trace.tennessee.edu/utk_agexcomhort/6.

World Bank. 2021. "Governance in Irrigation and Drainage Appendix 2: Case Descriptions." World Bank.

World Bank 2025. "Implementation Completion and Results Report." Report No. ICR00063, Agriculture and Food, South Asia, World Bank. https://documents1.worldbank.org/curated/en/099021125130542105/pdf/BOSIB-16984764-2911-4c57-8231-2c3832b88d31.pdf.

CHAPTER 5

Conclusion

Agricultural water management (AWM) lies at the intersection of strategic policy objectives related to people, prosperity, and planet. All three need to be duly considered in designing the AWM solutions of the twenty-first century. Context matters: Objectives will vary from place to place depending on production systems, trade options, water balance, and other factors. Solutions need to be tailored to each situation or risk failure.

Paul Harvey is widely attributed to have once said, "Man—despite his artistic pretensions, his sophistication and his many accomplishments—owes his existence to a 6-inch layer of topsoil and the fact that it rains." The fact is, however, that very often it doesn't rain, and 6 inches of soil holds very little water indeed. Possibly one of humanity's great accomplishments, therefore, will be managing the existential risk—with only 11 inches of freshwater to go around—of nourishing 10 billion people while sustaining a flourishing planet.

Modest adjustments in water use will not suffice to meet current or future food demands, which are projected to increase by up to 56 percent between 2010 and 2050 (Van Dijk et al. 2021). To accelerate inclusive, sustainable, farmer-led irrigation, we must act now to strengthen the enabling environment, reduce barriers for farmers, and realign the incentives of government, the private sector, and financial actors. This means we must do the following:

- Foster better links across the ecosystem, ensuring that farmers, suppliers, financial institutions, and agribusinesses are aligned, coordinated, and equipped to respond to each other's needs. Strengthening these links unlocks technology adoption, market access, and financing opportunities.
- Shift from building infrastructure to delivering sustainable water services. Investments should be guided by governance, incentives, and financing arrangements that support long-term performance, accountability, and financial viability—ensuring that systems deliver reliable services over time, not just physical assets. Investing in the "soft infrastructure"—knowledge exchange, communication outreach, digital

tools, and capacity building—has proved to be central to successful development outcomes.

- Adopt catalytic financing instruments, such as credit guarantees, blended finance, and risk-sharing mechanisms, to lower barriers, mobilize private capital, and enable farmers to invest with confidence. Redirecting a portion of current spending on agricultural support—combined with regulatory reform, blended finance, and public-private partnerships—can crowd in private capital and support financially sustainable, water-efficient service delivery.
- Use water availability and food trade patterns to guide policy choices. Governments should move beyond one-size-fits-all approaches and use local evidence on water stress, food deficits, and trade exposure to decide where to expand, rebalance, or limit agricultural water use—aligning production choices with local water realities and global markets, and making trade-offs explicit and accountable with data.

AWM spans a continuum—from farming that relies entirely on rainfall to systems that use irrigation to supplement it. Strategic, well-timed use of irrigation can stabilize production, sharply raise yields, and make rainfall more productive. Investments in AWM can generate large employment gains, and more productive water systems can reduce pressure on land and ecosystems. This report provides examples of how countries have achieved AWM transformation, including innovations that have worked and actions that countries have taken. Using this information, and the water-food nexus framework, policy makers can chart their own journeys toward improved AWM policies and results—sustainably nourishing a flourishing planet.

Reference

Van Dijk, Michiel, Tom Morley, Marie Luise Rau, and Yashar Saghai. 2021. "A Meta-Analysis of Projected Global Food Demand and Population at Risk of Hunger for the Period 2010–2050." *Nature Food* 2 (7): 494–501.

www.ingramcontent.com/pod-product-compliance
Lightning Source LLC
LaVergne TN
LVHW070955210926
848482LV00002B/218

* 9 7 8 1 4 6 4 8 0 7 8 2 4 *